以人为本的城市设计系列丛书

本书在完成过程中受到"江苏高校优势学科建设工程三期项目"……地
重划带动城市老工业区公共空间系统的更新改……。

People-oriented Urban Design
Inheritance and Transformation

以人为本的城市设计
传承与变革

段进　易鑫　等　著

东南大学出版社
·南京·

内 容 提 要

近年来，我国采取以人为本的发展方式，城市发展呈现出多样化的特征。以人为本的城市设计在技术上更为复杂，所需要的时间更多、成本更高，人们必须与当地居民进行更深入的互动，并充分反映当地的发展需求。各个城市面临的具体问题不尽相同，城市设计的行动框架也千差万别，因此有必要探索创新性的策略。

本书所载的11篇论文，是2019年第一届"以人为本的城市设计"国际会议所征集的优秀论文。在各个篇章中，各位作者结合不同国家城市的具体情况进行了详尽的阐述，并结合具体的城市设计案例，对以人为本的城市设计的实践、策略与政策进行了讨论。本书既可作为城市设计、国土空间规划、风景园林史等领域的研究资料，又可作为高等学校和社会各界人士了解城市设计发展变迁的参考用书和读物。

图书在版编目(CIP)数据

以人为本的城市设计：传承与变革／段进等著.—南京：东南大学出版社，2021.11（2024.5重印）
（以人为本的城市设计系列丛书／段进主编）
ISBN 978-7-5641-9784-1

Ⅰ.①以… Ⅱ.①段… Ⅲ.①城市规划—研究—中国 Ⅳ.①TU984.2

中国版本图书馆CIP数据核字(2021)第231314号

责任编辑：丁丁　责任校对：张万莹　封面设计：王玥　责任印制：周荣虎

以人为本的城市设计：传承与变革
Yirenweiben De Chengshi Sheji：Chuancheng Yu Biange

著　　者：段 进 易 鑫 等
出版发行：东南大学出版社
社　　址：南京四牌楼2号　邮编：210096　电话：025-83793330
网　　址：http://www.seupress.com
电子邮件：press@seupress.com
经　　销：全国各地新华书店
印　　刷：江苏凤凰数码印务有限公司
开　　本：787mm×1 092mm　1/16
印　　张：12.5
字　　数：296千字
版　　次：2021年11月第1版
印　　次：2024年5月第2次印刷
书　　号：ISBN 978-7-5641-9784-1
定　　价：78.00元

序　言

改革开放以来，中国经历了一个史无前例的快速城镇化进程，2019 年起城镇化率已突破 60%。从城市发展宏观背景来看，我国的城市发展已经进入以快速发展与结构性调整并行互动为特征的城镇化中后期阶段，正逐步由增量为主的规模外延扩张向增量存量结合并逐渐以存量为主的城市内涵式更新转型发展，此时，城市设计正成为这一转型发展的重要支撑，关于它的探讨也持续升温。《国家新型城镇化规划(2014—2020 年)》根据世界城镇化发展普遍规律和我国发展现状，明确提出要全面提升城市居民的生活质量，完善城市功能，提升公共服务水平和生态环境质量。

与西方国家的城镇化经历了多个世纪不同，中国的快速城镇化是在近几十年里发生的。面对这种快速发展的局面，人们需要更加系统地总结这个过程，基于历史的视角来把握各方面的经验。在当前社会、经济与文化要素急速转型的背景下，中国城市的空间结构、历史、文化等要素都面临转型与重构的要求。如何在尽可能尊重现有空间肌理的同时，发掘并释放城市空间新的活力，为社会、经济和文化的需求提供充满价值和吸引力的城市空间框架，无论在战略层面，还是在日常性的规划实践层面，都是无可回避的基本命题。

城市设计的发展向来都与国际经验密切相关，无论是历史上的还是今天的城市设计实践和构想，都是基于当地情况所做出的创造性活动。向国际学习，在本土行动！这里必须指出的是，城市设计不仅仅是寻找新的空间形式，城市设计本身必须综合社会各方面的诉求，把社会的整体需求、经济发展和尊重历史文化等问题放在一起综合考虑。对于城市设计也不能仅仅从学术角度思考，而是要始终应对当前的迫切任务。“以人为本的城市设计”国际会议提供了一个中外交流的重要平台，旨在探讨城市设计在世界范围内的重大意义，这也是我们主办本次会议的初衷所在。

在可持续发展背景下，全球正兴起重新关注城市设计的潮流。本书正是基于“以人为本的城市设计”国际会议交流而汇集的成果，入编论文的各位作者探讨了城市设计的前沿发展及一系列重要问题，从不同的学术或者实践视角开展了卓有成效的研究并取得了成果。本书希望激发大家的讨论，促进广泛的交流，同时加强城市设计师在追求可持续城市发展中的责任意识。我相信，本书能够为城市设计的转型提供学术导向性帮助，使这一非常重要的专业工具向着兼具经济活力、环境安全且社会包容性的模式发展。

王建国

东南大学教授，中国工程院院士

目　录

第三部分　城市设计与更新实践

第四部分　作者简介

第一部分

主旨论述

中国城市特色危机的人本性思考与规划应对
——武汉总体城市设计的探索

Humanism thinking and planning response to the Chinese urban identity crisis: A case study of Wuhan Comprehensive Urban Design

段 进 兰文龙 邵润青
Duan Jin Lan Wenlong Shao Runqing

摘 要:文章针对中国快速城镇化进程中出现的特色危机现象,运用 VOSviewer 软件梳理了近 30 年中国的城市特色研究,概括出"景观风貌""历史文化"和"产业经济"三个基本范式,指出人本性缺失是这些研究未能解决城市特色问题的根源。在系统阐释城市特色人本性内涵的基础上,提出城市特色研究在对象、主体和方法上的重大转向,并从内容、指标和结果判断三个方面构建出全新的城市特色评价体系,结合武汉总体城市设计案例,通过"描述—解释—评价—设计"的系统贯通,探索了中国城市特色危机的规划应对思路。

关键词:城市特色;特色危机;人本性;城市特色评价;城市设计

Abstract: Aiming at the identity crisis phenomenon appearing in Chinese rapid urbanization process, based on the analysis of Chinese urban characteristics studies in the past 30 years via VOSviewer, three research paradigms are summarized. There are (1) from the landscape perspective, (2) from the history and culture perspective, and (3) from the industry and economy perspective. It is pointed out that the lack of humanism is the main reason of these studies' failure to solve the characteristic issue. Furthermore, on the basis of systematically explaining the connotation of humanism with urban identity, this paper puts forward the major changes in the research object, subject and method. And an urban characteristic evaluation system, including content, index and result judgement, is constructed. Finally, based on the case of Wuhan Comprehensive Urban Design Project, through the description-evaluation-design-explanation framework, the planning strategy of dealing the Chinese urban identity crisis is explored.

Key words: Urban identity; Identity crisis; Humanism; Urban identity evaluation; Urban design

自1980年代以来，中国步入快速城镇化进程，城市面貌日新月异，取得了令世人瞩目的成就。但与此同时，“千城一面”“新、奇、怪”等城市特色问题愈发严峻，引发了社会各界的广泛关注与思考。

近年来，塑造城市特色风貌逐渐成为城市治理领域的重点议题。2013年中央城镇化工作会议提出：“要依托现有山水脉络等独特风光，让城市融入大自然，让居民望得见山、看得见水、记得住乡愁。”2015年中央城市工作会议强调：“不断提升城市环境质量、人民生活质量、城市竞争力，建设和谐宜居、富有活力、各具特色的现代化城市。”2016年《中共中央国务院关于进一步加强城市规划建设管理工作的若干意见》再次重申：鼓励开展城市设计工作，统筹城市建筑布局，协调城市景观风貌，体现城市地域特征、民族特色和时代风貌。

在中国全面建成小康社会、未来要更好满足人民美好生活需要的时代背景下，如何在规划建设中体现地域特色与时代特征，实现人民群众的高品质生活和国土空间的高质量发展，是新时期城市工作的重点，也是建设“美丽中国”的核心要求。

1　近30年中国城市特色研究的三个基本范式

针对城市特色这一议题，中国学术界进行了长达30年的研究探讨。为了进一步梳理其脉络，厘清科学问题，课题组运用VOSviewer软件，选取CNKI中国知网作为数据来源，以“城市特色”为主题词，检索1989年至2019年间的4 446篇文章并对其进行量化分析。研究发现，这些成果呈现出较为明显的“领域分异”特征，可概括为“景观风貌”“历史文化”和“产业经济”范式下的三个研究聚类(图1)。

1.1　景观风貌范式

一些学者从景观风貌视角强调一个城市的特色是其鲜明(唯一性)外在特征的集中体现[1]。他们立足城市形态演化中所形成的独特禀赋[2]，关注空间要素之间的“特殊性”与“普遍性”，认为城市特色通过“特殊性要素”与“普遍性要素”的辩证关系而呈现[3]。在此范式下，特色危机的优化路径在于解决空间要素个体与群体、虚与实、新与旧的相互关系[4]，如林奇(K. Lynch)将城市意象归纳为经典的路径、边缘、区域、节点和标志五要素[5]，齐康院士提出的轴、核、群、架、皮[6]等，实现了城市特色的“结构性”勾勒。但是，这些研究因为徘徊在以城市视觉形象为基础的陈旧表面[7]，采用建设地标、控制环境色彩、复制古建筑等治标性措施[8]，造成风格杂糅、拼贴感严重[9]等视觉环境问题，受到了一定程度的质疑。包括荷兰著名建筑师库哈斯(R. Koolhaas)等在内的一些学者就结合21世纪初中国深圳等地的建设，进行了“多即是多”的激烈讨论[10]。

1.2　历史文化范式

另一些学者认为城市特色实际包含“风”和“貌”两个部分，且风是貌的内在支撑，故将研究重心转移到历史文化方面[4]。他们认为城市特色危机的根源是风的问题，即城市建设急功近利，忽视对历史文脉的尊重和保护，使历史空间在快速发展中被商业开发蚕食、打破，取而代之

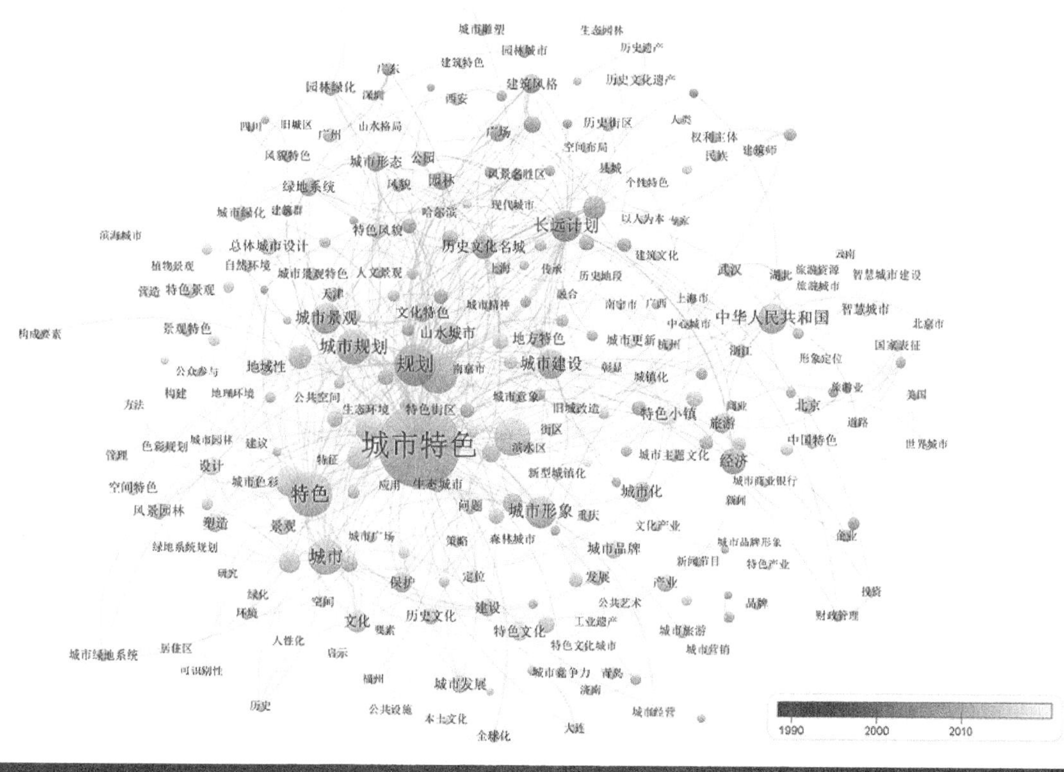

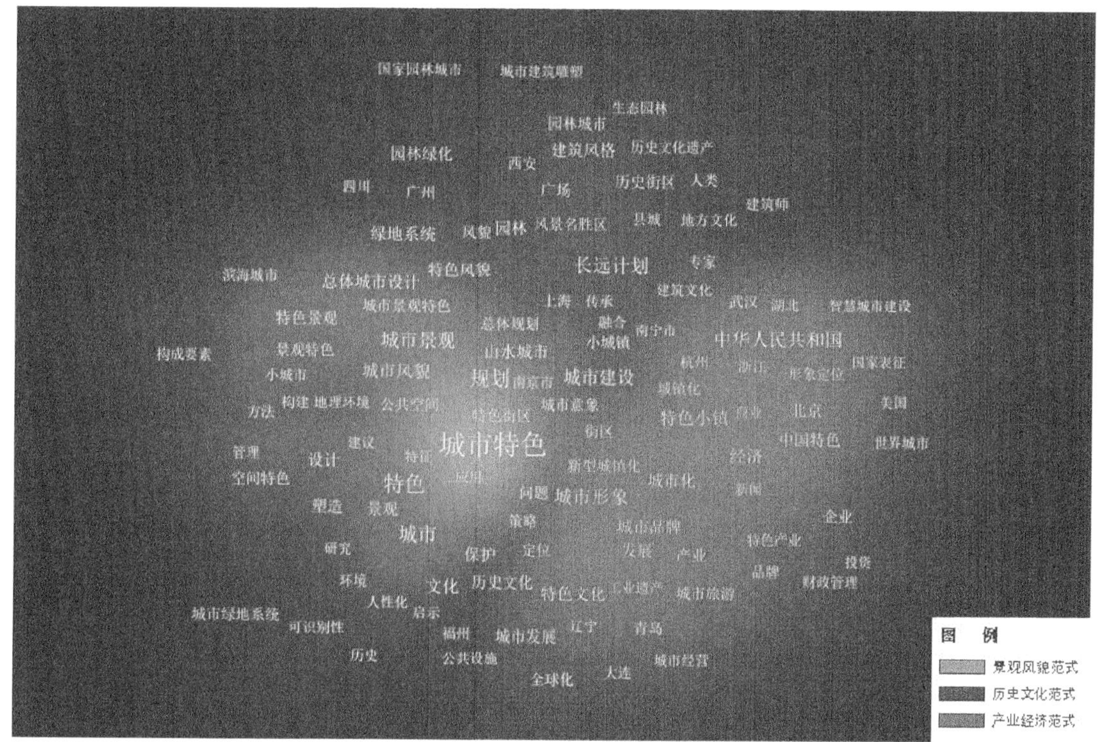

图 1　基于 VOSviewer 软件的城市特色研究聚类分析

的是模式化的现代工业产品[3]。在应对思路上，他们更偏重对地域文化、历史文脉和建筑风貌的挖掘与再创造[11]。但也有学者指出，这种范式因过度依赖设计师的创作能力而具有偶然性，特别是在“西方文化先进论”思想影响下存在巨大风险。倘若对本土文化缺乏了解，极易造成历史空间的随意曲解、嫁接，甚至推倒重来[12]，反而导向一种“空有其形而无其神韵”的距离感以及快速变革中的不适感[13]。

1.3 产业经济范式

还有一些学者突破了传统的城市形态研究范畴，从更为宏观的产业经济视角对城市特色的“资源”属性进行解读。他们认定是城市发展战略的偏差导致了特色危机——以短期经济效益为目标，盲目复刻甚至照搬其他城市经验[7]，而对自身的资源条件和竞争优势意识淡化，忽视了城市发展的质量和城市特色的营造[3]。加之经济全球化背景下，规划设计师、建筑材料、建造技术，甚至资本的快速流动，更助推了特色危机在世界范围内的蔓延态势[14]。针对此，江泓、余柏椿等学者建设性提出，城市特色作为一种资源具有商品性，这既是特色问题的起因(空间商品化使城市建设对地域文化难以及时响应)，也是题眼所在，通过较为成熟的区域比较优势分析方法，可深入挖掘地方资源禀赋，使其获得城市发展的内生动力和经济效益[3, 15, 16]。然而这种“资源先决论”观点有两个逻辑死结：一方面将不具备先天条件的广大城市排除在外；另一方面忽视了人在城市特色形成和发展过程中的能动作用[17]，同样面临不少质疑。

通过分析可以看出，城市特色一直是学界研究的一个热点，但特色危机的现实状况却未见好转[8]。笔者认为，一个重要的原因是这些研究大多聚焦于环境客体，以“精英”视角对城市特色的各类载体进行创造性形态组织与设计，完全忽略了普通居民、游客等具体行为主体的真实感受。因为广大市民才是城市特色认知的主体，城市特色的塑造就是为了满足人民日益增长的美好生活需要[14]，只有强化具体人的主体性，准确把握城市特色的人本性内涵，才能从根源上实现中国城市特色危机的研究破题。

2 城市特色研究的“人本性”转向

从人本性视角，可将城市特色解读为公众基于城市空间在形态上的特殊性，通过各式各样的感知和体验活动，在充分的环境行为互动过程中形成的集体环境意象(图 2)。

首先，城市空间在形态上的唯一性和排他性是认知城市特色的前提。关于这一点，林奇[5]、罗西(A. Rossi)[18]、诺伯格-舒尔兹(C. Norberg-Schulz)[19]等学者已做过大量研究和充分佐证。其次，公众在城市空间中的活动情况决定了城市特色认知的深度。作为市民日常生活容器的城市空间，只有与市民活动发生关联、符合人们感知与体验需求时，才可能成为城市特色。最后，城市空间赋予公众意义是城市特色认知的结果。对于城市特色，每个人都有不同的理解，今天我们所讨论的城市特色，实则是一种普遍化的心理意象，即社会价值观在城市空间上的反应[8]，而非环境行为之间普通的物理关系[20]。

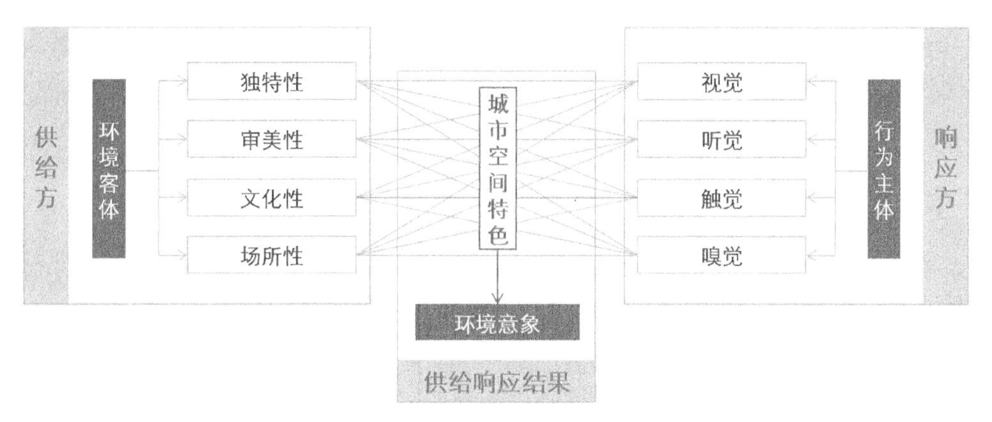

图 2　从人本性视角解读城市特色的形成机制

在此认知下，城市特色的研究对象、主体和方法将发生重大转向。首先，城市特色不再是通常意义上的某一座山、某一条河或是某一幢建筑，而是这些物化形态经过人的感知与体验后形成的、对于空间的特殊理解和认知，即从"空间场所"到"公众行为"再到"环境意象"的复杂过程。其次，城市特色的研究不再局限于"精英"族群，而是涵盖城市管理者、专家、普通市民、游客、开发商、规划师等所有利益相关者(Stakeholder)的多元主体。相应的，城市特色研究脱离了以主观定性为主的专家法，而采用在精确性和严谨性方面更为突出的心理物理学派方法，利用其主客观相结合的特性，在环境客体与行为主体之间进行关联量化分析，从而获取影响城市特色公众感知的客观因子(表 1)。

表 1　城市特色研究的"人本性"转向

类型	传统城市特色研究	人本性城市特色研究
研究对象	物化形态：一个城市鲜明的外在形态特征(资源)，如标志性的山体、水体、建(构)筑物等，具有排他性和唯一性	城市意象：环境行为互动过程中形成的集体环境意象，并非具体的"物"，而是"物"与"人"的对应关系，具有可感知性
研究主体	精英：城市管理者、专家和规划师等专业人群，具有将城市视为"艺术品"的倾向	公众：市民、游客、开发商、城市管理者、专家、规划师等所有利益相关者，城市是其日常生活的环境
研究方法	专家学派方法：依据少数专家的知识和经验对城市特色问题进行分析和预测，以定性为主，可操作性和适应性强	心理物理学派方法：运用量化手段在城市特色客观属性与公众行为之间进行关联性分析，找出影响公众行为的客观因子，主客观相结合，具有精确性和严密性

3　基于人本理念的城市特色评价方法

笔者认为，基于人本理念客观呈现特色载体与公众之间的关系，是城市特色规划设计实践的必要前提，因而构建出一套全新的城市特色评价方法。其与常规方法的区别在于：在内容上更关注公众对特色载体的视觉感知与活动体验，而不仅仅将城市特色视为一种资源；在指标上更突出感知类与非感知类指标在环境供给与行为响应方面的互动校核，减少单一维度评价可能带来的偏颇；在结果评判上更强调“描述—解释—评价—设计”系统[16]的严谨性，避免评价工作成为可有可无的参照性内容。本文借助武汉总体城市设计案例①进行详细说明。

3.1 评价内容

常规的资源竞争力评价方法在测度特色价值方面已较为成熟，本研究借助该方法筛选相应的特色载体作为评价对象，同时根据人对城市特色感知途径的差异，构建出“视觉感知＋活动体验”两大评估维度，并从环境客体和行为主体的不同角度反映环境供给能力和行为响应程度，从而全方位呈现特色载体与公众之间的相互关系。

在武汉案例中，城市特色评价主要包含三个步骤两个层面内容：首先是价值层面的特色资源评价——通过自然山水、历史人文、都市发展等三方面资源的挖掘与竞争力分析，明确“两江交汇”“百湖之市”“十字山水”等九类空间特色类型，继而划定六十二片城市特色空间载体作为评价对象（图3）；其次是感知层面的单一特色载体评价——从公众视觉感知与活动体验两个

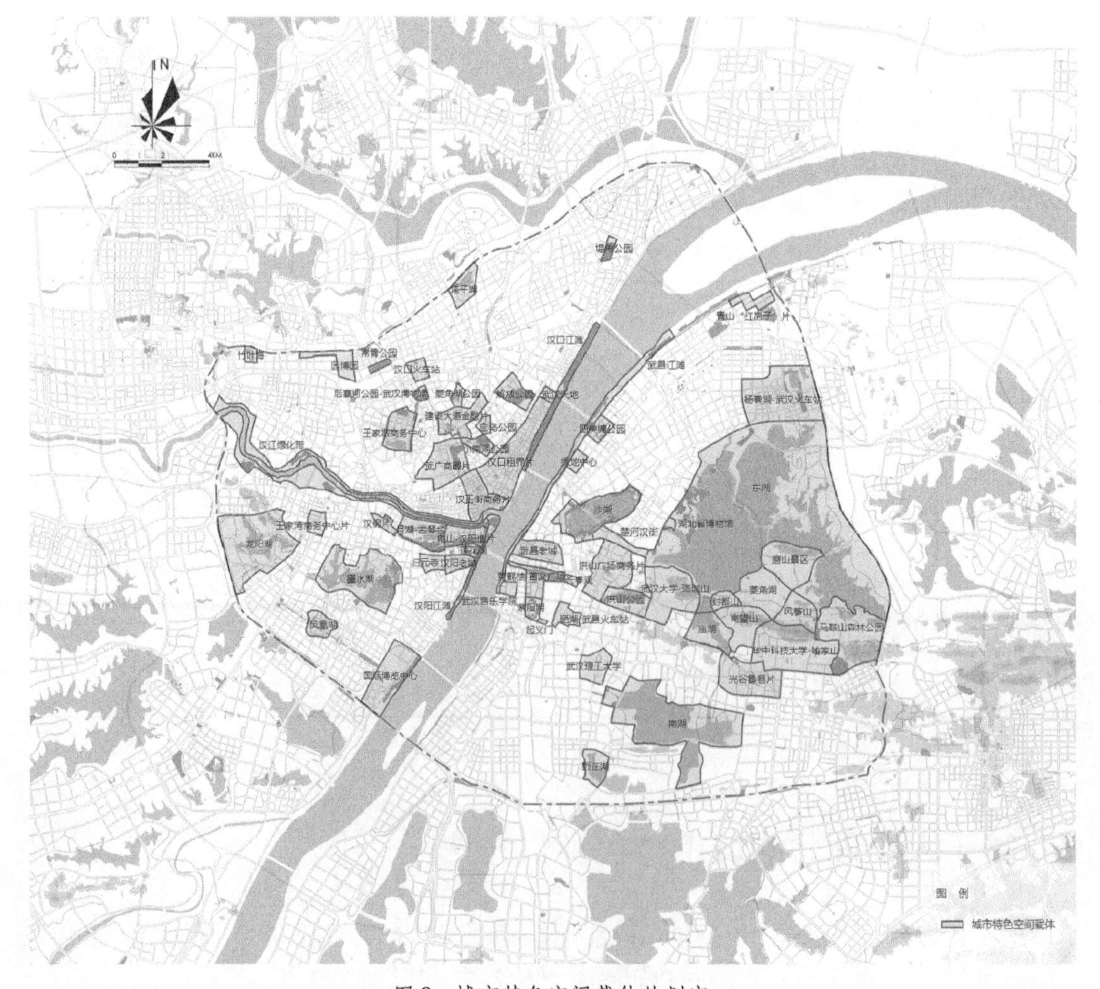

图3　城市特色空间载体的划定

维度，结合可识别性、审美属性、可意性、活力和活动支持五个影响因子对单一特色载体进行分析，明确其在环境供给与行为响应中存在的具体问题；最后是感知层面的整体特色评价——单一特色载体评价结论按照类型归属被组织到城市整体特色系统的描述与研判中，形成城市设计问题矩阵，并作为城市景观风貌体系与公共空间体系导控的直接依据。

3.2 评价指标

按照从视觉感知到活动体验、从表层形态到深层结构的顺序，明确视觉敏感度、美景度、可意度、活力度和活动支持度五个一级评价指标及其相应的二级指标，形成层级化的城市特色评价指标体系(表 2)。同时引入可感知指标概念，将所有指标分为感知与非感知两种类型，从公众与特色载体的不同角度测度其互动效果。例如，美景度是反映公众关于特色载体审美态度的感知类指标，活动支持度(如设施密度、功能混合度等)则是测定特色载体支撑公众活动水平的非感知类指标。此外，研究指标涉及多种数据源，将其放在同一个评估体系中，需转化为可统一比较、处理的数据，常用的方法包括德尔菲法、层次分析法和度量衡法等。

表 2　城市特色评价的指标体系

评价维度	一级评价指标	指标类型	二级评价指标	指标类型
视觉维度	视觉敏感度	感知类指标	距离敏感度	非感知类指标
			出现概率敏感度	非感知类指标
			形体醒目度	感知类指标
			色彩醒目度	感知类指标
	美景度	感知类指标	绿色比	非感知类指标
			水域比	非感知类指标
			天空开阔度	非感知类指标
			色彩数量	非感知类指标
			色泽	非感知类指标
			建筑质量	感知类指标
			历史要素	非感知类指标
			和谐度	感知类指标
			韵律感	感知类指标
			视觉污染	非感知类指标
	可意度	感知类指标	形态简单性	非感知类指标
			连续性	非感知类指标
			统领性	感知类指标
			全览性	感知类指标
			文脉	感知类指标

（续表）

评价维度	一级评价指标	指标类型	二级评价指标	指标类型
活动维度	活力度	感知类指标	人群多样性	非感知类指标
			人群密度	非感知类指标
			活动多样性	非感知类指标
			活动时间	非感知类指标
	活动支持度	非感知类指标	区位	非感知类指标
			交通可达性	非感知类指标
			连通度	感知类指标
			公共性	感知类指标
			功能混合度	非感知类指标
			设施密度	非感知类指标

在武汉案例中，利用非感知类指标直接测度特色载体的环境供给能力，如用公交站点距离反映特色载体的可达性、用对象与周边要素的形体差异反映特色载体的视觉敏感度等。利用感知类指标直接测度特色载体的公众响应程度，如用百度热力图反映特色载体中的活跃人群数量、用“我心中的武汉——武汉总体城市设计公众调查”数据反映公众对于特色载体意义的认知情况。同时尽可能通过非感知类与感知类指标的相互转化，实现形态特征和人本绩效的完整阐释。如对美景度的评价，采用 2004—2015 年间武汉主城区 flickr 照片频次以及 2018 年携程网景点评分等大数据反映特色载体的公众审美态度，同时对特色载体的绿化比、水面比、天空开阔度等客观因子进行量化，通过相关性分析找到影响公众审美的客观因子，建立环境供给与行为响应之间的因果链带。

3.3 评价结果判断

利用资源竞争力、环境供给能力和行为响应程度相结合的评估内容，可得出各特色载体和城市整体特色的单因子指标、复合因子指标得分与总得分(图 4、图 5、图 6)。通过对综合评估结果的分析，各个维度内容的判读，以及重要单因子的考证，形成较为全面的评估结果。其中，重点关注供给与响应之间是否存在偏差，并以此作为规划设计实践的着力点。

在武汉案例中，通过以上方法研究发现，“百湖之市”是武汉具有显著竞争优势的特色资源，但在九类基础空间特色的视觉感知与活动体验评价中却排名末尾，显现出资源竞争力与特色感知效果之间的错位。具体而言，“百湖之市”特色载体的连通性、视觉敏感度、设施密度等供给能力指标得分较低，且活力度、可意度等衡量公众响应程度的指标得分也远低于平均水平。针对此，课题组从促进城市特色公众感知、实现特色载体与公众之间多层次耦合的角度对评价结果进行反馈，于宏观城市、中观社区、微观场所三个层面分别提出了“9 + 6”滨水公共中心体系(对应活力度、设施密度)、江湖连通(对应连通性)(图 7)、蓝绿纤维织补(对应视觉敏感度)等规划设计策略，实现了“描述—解释—评价—设计”系统的贯通。

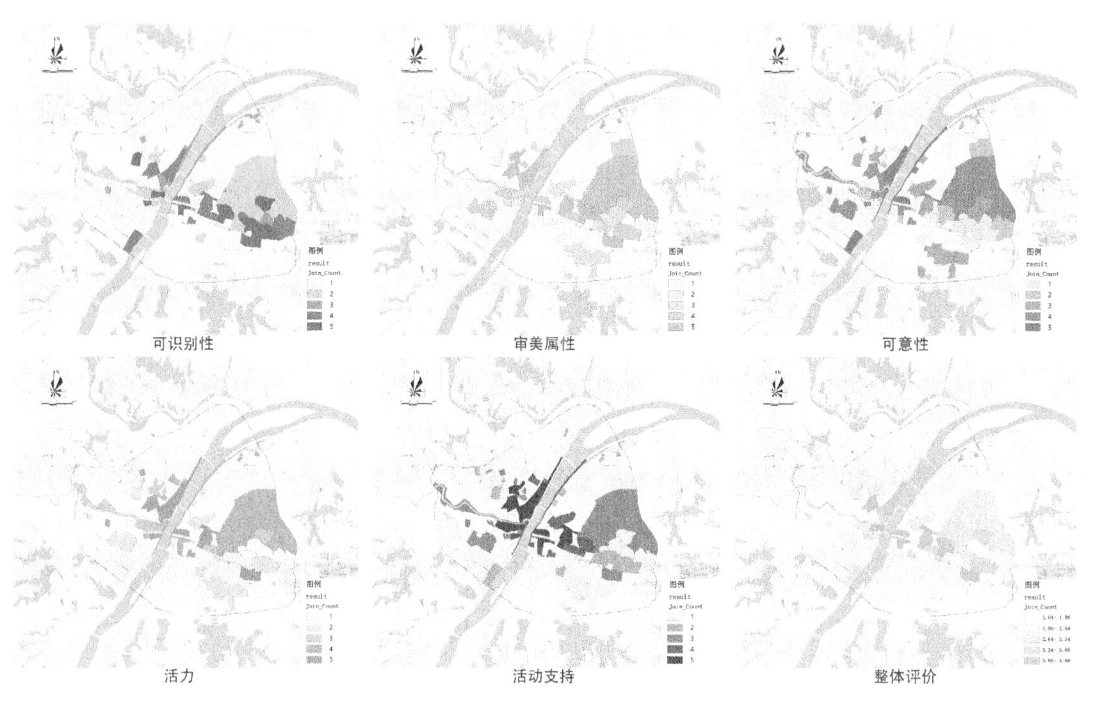

图 4　单一特色载体的单因子指标得分及复合因子得分

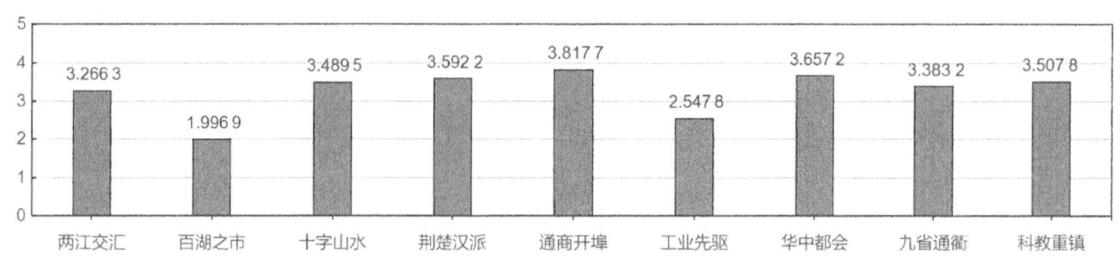

图 5　九类基础空间特色的总得分

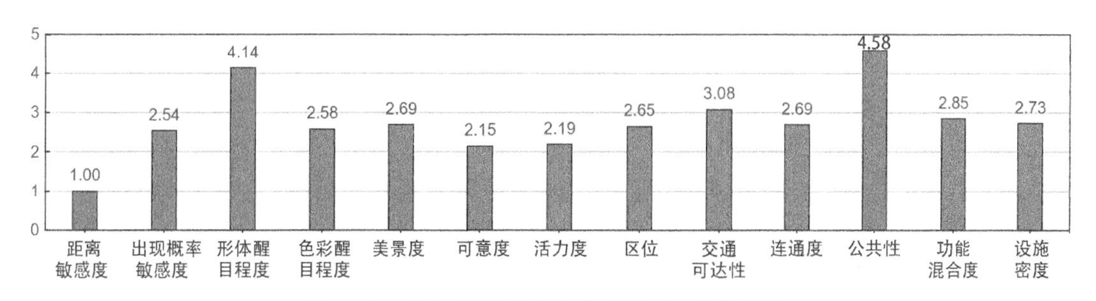

图 6　“百湖之市”类特色载体的单因子指标得分

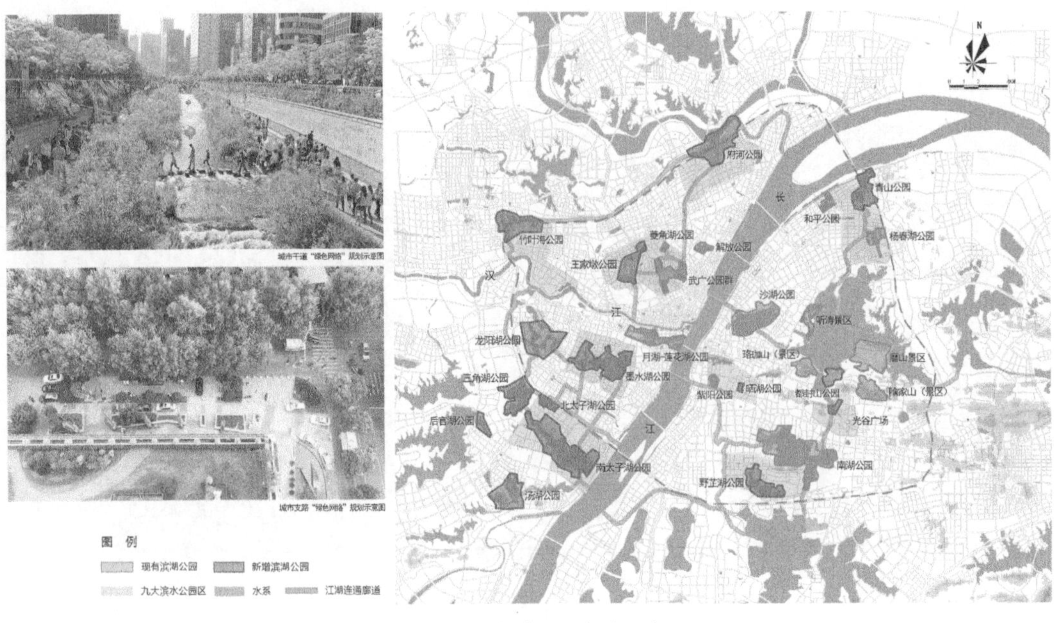

图 7 “江湖连通”计划示意

4 结语

过去三十年间，中国学者在城市特色提炼和创造方面累积了相当的研究成果，但在“什么才是当地的特色”“特色资源与公众的关系”方面却存在一些误区，各地屡屡出现“仿古一条街”“欧式一条街”[21]等实践恶果，不得不说是一个遗憾。

当下的城市特色风貌塑造工作，需要从价值观和方法论两方面同时入手，不仅关注表层上的物质空间形态塑造，而且要深深植根于人民群众对城市特色的需求和向往，运用合理的评价手段建立起物质空间与公众意象之间的因果链带，再因循展开规划与设计。在这一方面，武汉总体城市设计案例率先进行了探索，通过基于人本理念的城市特色评价方法运用，使兼顾揭示性和指导性的评价工作，变成了真正扭转特色危机现象的适用且有效杠杆。该成果被纳入《武汉市城市总体规划(2017—2035 年)》相应章节，在城市整体发展目标、空间格局、用地和重大设施布局等方面发挥了重要作用[22]，并作为探索特大城市总规编制中运用城市设计方法的案例，获 2019 年度全国优秀城市规划设计奖一等奖②。

值得注意的是，城市特色评价是一个持续的过程，在规划编制和城市建设的各个阶段都应进行基于人本理念的评价工作，自上而下与自下而上相结合，以确保及时掌握公众与特色载体之间关系的变化情况，进行“描述—解释—评价—设计”系统的适时调整与动态更新。

注释：

① 武汉总体城市设计课题由东南大学城市规划设计研究院段进团队和中国城市规划设计研究院朱子瑜团队合作完成。

② 该项目获奖人包括：段进、朱子瑜、涂剑、邵润青、陈振羽、姜涛、兰文龙、马璇、李丹哲、王里漾、刘珺、

胡忆东、郭宜仪、孙晓敏、徐晨慧。

参考文献

[1] 王浩,王亚军.城市绿地系统规划塑造城市特色[J].中国园林,2007,23(9):90-94.

[2] 郑浩,王丰,宣甲,等.山水城市总体城市设计的探索:以台州市中心城区为例[J].城市规划,2020,44(S1):106-119.

[3] 江泓,张四维.生产、复制与特色消亡:"空间生产"视角下的城市特色危机[J].城市规划学刊,2009(4):40-45.

[4] 吕斌,杨保军,张泉,等.城镇特色风貌传承和塑造的困与惑[J].城市规划,2019,43(3):59-66.

[5] 林奇.城市意象[M].方益萍,何晓军,译.北京:华夏出版社,2001.

[6] 齐康.城市建筑[M].南京:东南大学出版社,2001.

[7] 申绍杰.批评的反省和辨析:千城一面再认识[J].建筑学报,2013(6):96-98.

[8] 薛立新,龙彬.从社会发展角度认识城市特色危机[J].城市规划,2012,36(4):37-41.

[9] 陈雨,伍敏,刘中元,等.历史文化城市空间特色规划编制方法探索:以黄山市实践为例[J].城市规划学刊,2017(S2):92-97.

[10] Koolhaas R. Junkspace[J]. October, 2002(100):175-190.

[11] 张巍,蒋朝晖,魏钢.古寨新城小城大事:以《北川新县城总体城市设计》为例的城市特色空间尺度研究[J].城市规划,2011,35(S2):43-46.

[12] 阳建强.城市历史环境和传统风貌的保护[J].上海城市规划,2015(5):18-22.

[13] 黄宗仪.都市空间的生产:全球化的上海[J].台湾社会研究季刊,2004(53):61-83.

[14] 袁锦富,司马晓,张京祥,等.城乡特色危机与规划应对[J].城市规划,2018,42(2):34-41.

[15] 余柏椿.城镇特色资源先决论与评价方法[J].建筑学报,2003(11):66-68.

[16] 梁鹤年.再谈"城市人":以人为本的城镇化[J].城市规划,2014,38(9):64-75.

[17] Choi H S, Reeve A. Local identity in the form-production process, using as a case study the multifunctional administrative city project (Sejong) in South Korea[J]. Urban Design International, 2015,20(1):66-78.

[18] 罗西.城市建筑学[M].黄士钧,译.北京:中国建筑工业出版社,2006.

[19] 诺伯格-舒尔兹.场所精神:迈向建筑现象学[M].施植明,译.武汉:华中科技大学出版社,2010.

[20] 李斌.环境行为学的环境行为理论及其拓展[J].建筑学报,2008(2):30-33.

[21] 段进.城市空间特色的符号构成与认知:以南京市市民调查为实证[J].规划师,2002,18(1):73-75.

[22] 段进,兰文龙,邵润青,等.总体城市设计战略:一种新的城市设计类型:以武汉为例[J].城市规划,2018,42(z2):51-57.

在气候变暖的条件下为人民设计

Designing for people in a warming climate

［美国］乔纳森·巴奈特
Jonathan Barnett

多年以来,中国在经济和城市扩张方面取得了巨大成功。现在,中国政府已经越来越重视建成环境质量,以人民为本的城市设计政策实际上是对之前人们把效率和经济增长作为优先目标的修正。在这一背景下,人们有必要回顾过去,评估并改进既有的工作;同时还要展望未来,气候的重大变化已经开始,其后果将不可避免。在气候变化时期,为人民进行设计将人们的安全置于非常优先的地位。

生活在不断变暖的星球上,人们面临七类主要的威胁:(1)海平面上升和风暴潮,(2)河流洪水,(3)极端降雨事件,(4)威胁生命的高温,(5)干旱,(6)粮食短缺,(7)野火。直到最近,人们仍认为这些问题要到本世纪中叶甚至更晚的时候才会变得严重。但是,专家们低估了气候变化的速度。最近的报纸在头版头条记录了2019年9月一场7米高的风暴潮席卷了巴哈马群岛,而这距离高度脆弱的迈阿密海滩仅320千米,而且迈阿密海滩的海拔只有1.5米。在2019年异常高涨的潮汐之后,威尼斯是否能够承受反复的潮汐洪水正在成为一个问题。创纪录的河流洪水最近摧毁了许多地方,包括2019年的美国中西部地区。自2016年以来,得克萨斯州休斯敦附近的暴雨造成了500年一遇和200年一遇的洪水事件。2019年6月,创纪录的热浪席卷欧洲大部分地区。南非的开普敦在2018年几乎耗尽了饮用水。由于干旱和粮食短缺,从农作物歉收的乡村地区迁出的人口数量正在不断打破纪录,这其中包括逃离中美洲几个国家的人们,这给墨西哥和美国造成了难民危机。野火在2018年和2019年摧毁了加利福尼亚的部分地区,甚至在瑞典的北极圈地区都发生了类似事件。

北京大学俞孔坚教授及其同事李海龙、李迪华、乔青、奚雪松在2009年开展了一项名为“国土生态安全格局:再造秀美山川的空间战略”的研究。这项研究着眼于保护自然系统、生态修复的潜力以及中国各地适合建设的场所(图1)。

俞孔坚教授及其同事将保护和恢复自然系统的问题分为五种生态安全格局:水源涵养、洪水调蓄、沙漠化防治、水土保持和生物多样性保护。通过地理信息系统对每个类别中的四个或五个子系统进行了阐述,然后再通过地理信息系统将以上信息组合在一起,针对每个安全格局分别得到一个优化了这些因素的地图。然后,再通过计算机处理,将这五种安全格局综合在一起,以生成对整个国土的生态保护优化成果。这项开创性的工作是研究如何保护自然系统并与之和谐相处的环境背景和模型(图2)。

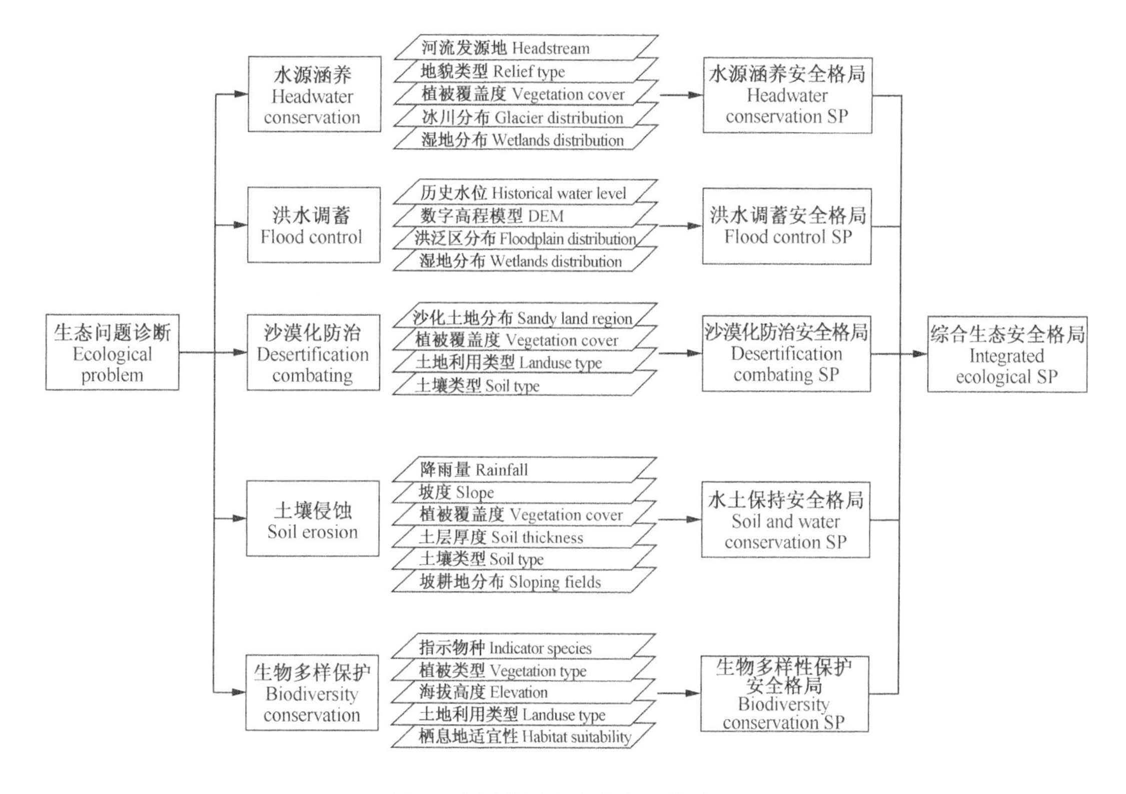

图 1　中国的国土生态安全格局*

Fig.1　China national ecological security patterns

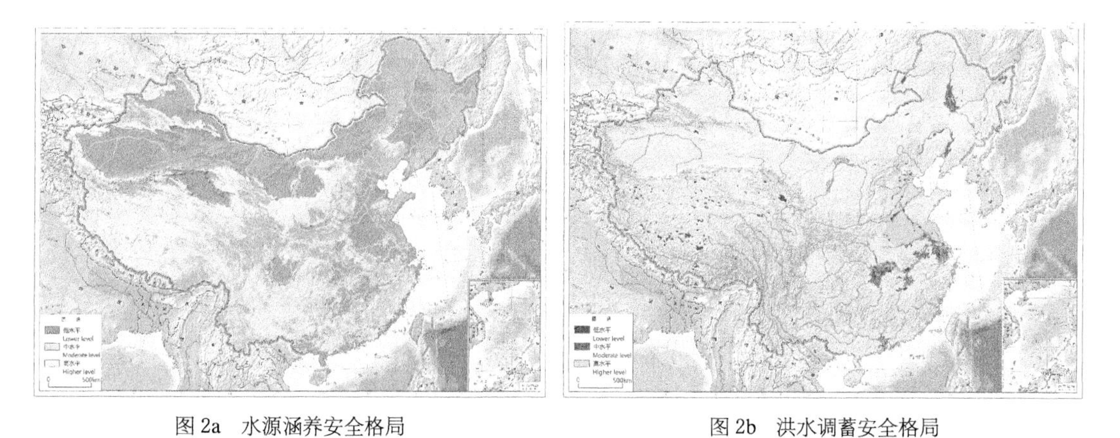

图 2a　水源涵养安全格局

Fig. 2a　Ecological security pattern for headwater conservation

图 2b　洪水调蓄安全格局

Fig. 2b　Ecological security pattern for flood control

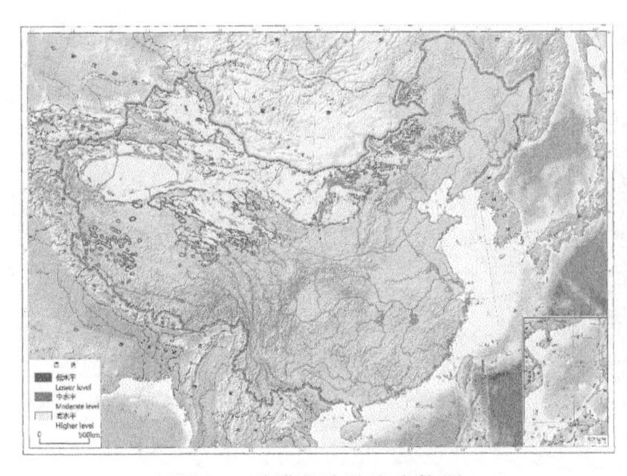

图 2c　沙漠化防治安全格局

Fig. 2c　Ecological security pattern for desertification combating

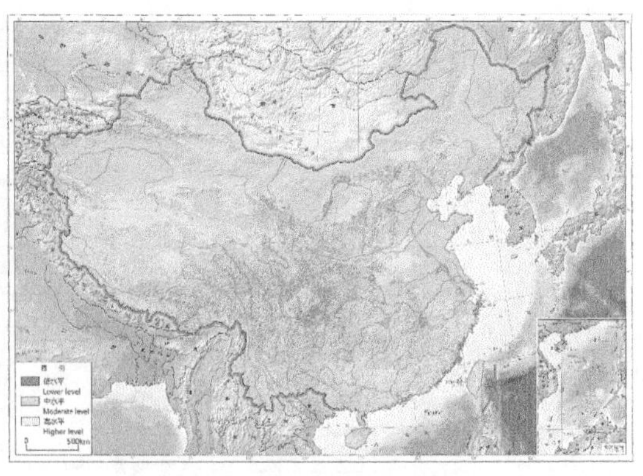

图 2d　水土保持安全格局

Fig. 2d　Ecological security pattern for soil and water conservation

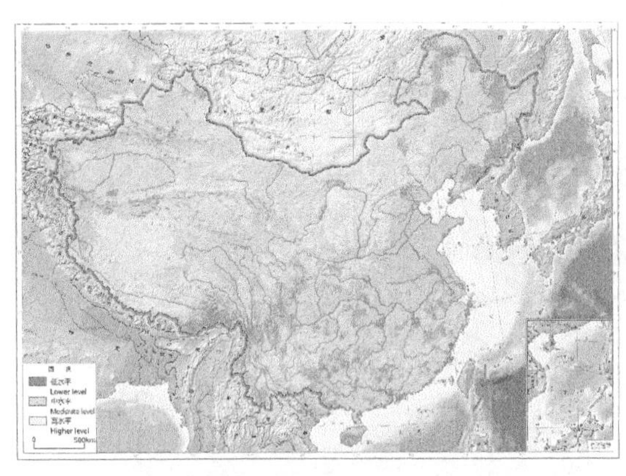

图 2e　生物多样性保护安全格局

Fig. 2e　Ecological security pattern for biodiversity protection

图 2f　综合生态安全格局

Fig. 2f　Integrated national ecological security pattern

图 2　生态安全格局分类

Fig.2　Classification of ecological security pattern

注：根据标准地图服务系统所提供地图[审图号：GS(2016)1609 号]改绘。

气候变化所释放的力量会影响到方方面面，包括在哪里建造、何时建造以及要建造什么的城市设计决策，而且这些内容不是要等到 2050 年或者 2030 年，而是应当立即做出决定。

不管未来如何采取措施限制全球变暖，大气中已经有足够的温室气体。在未来十年内，需要对设计和开发实践进行重大改变。到本世纪中叶，为了应对气候挑战，有必要发展出一种新的规划和城市设计实践。

在这个过渡过程中，确保人民的安全至关重要，这给规划师和城市设计师带来了许多挑战。

1 在哪里安全建造

应保留尽可能多的农田，而不仅仅是基本农田以增加粮食安全，还需要制定区域设计政策，以保护和恢复自然景观。采取此类政策的目的主要是存储降雨带来的径流，同时防止形成热岛。此外还需要进行景观的保护和重建工作，以增强所有自然系统抵御气候压力的能力。

例如高速公路和铁路等基础设施的新投资应该在可以远离洪水或野火的地方进行。而在预计存在风险的地方进行基础设施建设则需要加以保护或重新布线。这就需要通过区域设计方案来为这些情况做准备。

现有社区的扩展和新社区的选址，应在不受洪水和野火危害且不损害农业的地方建造。这仍然需要区域设计和规划来确定应进行开发的区位。

现有的沿海和沿江城市及其相关基础设施(包括高速公路、铁路和发电厂)将需要精心设计防护措施，以防御洪灾，并保卫现有城市和新建的建筑。在工程中需要针对这些设施提出最为合适的城市设计。

同样，需要按照区域设计所提出的要求，保护受到野火威胁的场所和基础设施，并且应该严格限制这些地方进行新的开发。

对于那些无法确保现有建筑物和基础设施免遭洪水或野火的地方，有必要从今天开始制定限制新建活动的规则，然后减少政府服务，再结合具体计划帮助人们和企业搬迁。这就需要重新设计，这些工作要从今天开始，或到 2050 年之前分阶段逐步撤离和清理。关于如何帮助人们应对生活中这些重大变化的决定将会非常困难。

在大尺度的城市设计方案中，可以确定需要从危险地区迁出人口的替代地点，如果满足安全要求，则优先改善被忽视和未得到充分利用的地区，而不是开发那些非建设的绿地。

无论是在新开发的还是在现有的城市化地区，土地使用和交通建设都应一体化进行，以减轻对于自然环境和农业用地的增长压力。最高强度的商业发展区位应集中在高铁站的周边地区。这些站点的交通连接良好，可以直达机场。居民区的设计应接驳通勤交通，快速公交系统(BRT 线路)可以在现有的高速公路走廊上开发，在这些地方进行轨道建设的投资成本较低。

2 何时制定安全措施

考虑到必要的建设工作可能要花费数十年的时间，因此需要开始为脆弱的城市地区制定保护工程的计划并提供资金。由于环境条件和气候变化的预测差异很大，因此需要设计具体的保护措施以满足特定区域的安全要求。

在禁止新建活动以后，还需要安排从危险地区撤离的时间表。要知道，这一过程可能还需要数十年的时间，并且替代地点应尽可能免受重大气候威胁的影响。

环境保护的基本原则：在充分利用现有城市化地区的潜力之前，不应将新的地区进行城市化。认识到这一原则将意味着城市增长管理方式的重大变化。新政策应为向那些被忽略和未充分开发的区域进行再投资，并设定时间表。

3 建造什么

为了减少导致全球变暖的排放,新的城市投资应该集中在步行、交通便利的零碳社区中进行,以减少能源和水资源的消耗,降低对汽车的需求。城市设计应包括废弃物回收利用管理,并确定使用太阳能和风能的区域。著名的例子包括斯德哥尔摩的哈默比城区(Hammarby Sjostad)和加拿大温哥华的福溪南地区(False Creek South)。改造现有社区以使其尽可能节能的设计也将是当务之急。这两个示例都包括一个面向社区的小型发电厂,该发电厂利用了来自浴室、洗碗机和洗衣机中使用的热水以及废水管中释放的热量,以减少发电所需的燃料量。当地社区收集的有机废物也可以为当地工厂提供生物燃料。

大规模和分布式的能源替代方案都应到位。在有土地供应的情况下,应在最合适的地方建造风力和太阳能发电场。在单个建筑物上布置太阳能和风能设施,可以减少传输带来的能量损失。现在已经拥有有效的电池存储技术,因此人们日夜均能利用可再生能源。

对于沿海大城市而言,防洪涌保护的例子包括俄罗斯圣彼得堡的屏障、伦敦的泰晤士河屏障等工程。圣彼得堡的屏障将涅瓦河的大部分河口与芬兰湾分开。为了让船舶通过,连贯的大坝会被两个可移动的闸门临时分成数段(参考鹿特丹的闸门),并设有6组防洪屏障,通常情况下允许河水正常流出,但在有洪水预警的时候可以关闭水闸(图3)。大坝顶部,在运输闸门

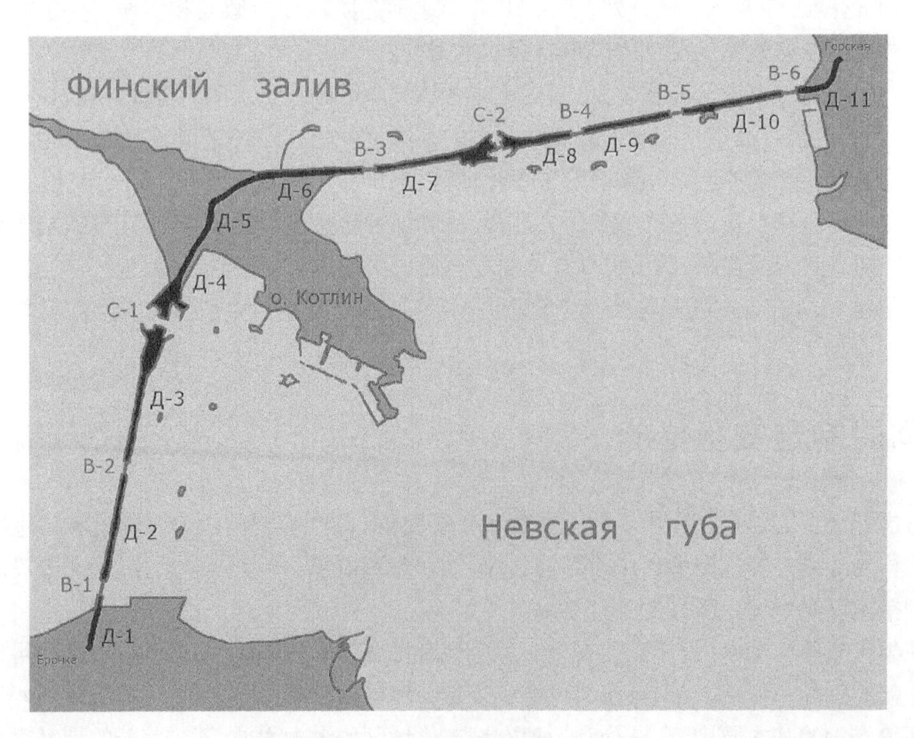

可以关闭该屏障,以防止波罗的海的洪水通过城市。正常情况下,闸门保持打开状态,以使河流通畅,船只可以进出。

图3 俄罗斯圣彼得堡的屏障计划

Fig.3 Plan of the barrier at St. Petersburg in Russia

防洪堤已并入圣彼得堡的环形道路。高速公路位于屏障的顶部。在入口处,这条路进入了河下的隧道。

图 4 屏障顶部的高速路

Fig.4 The highway running on the top of the barrier

的下方设有隧道,被作为整个城市环形道路的一部分(图 4)。人们在屏障内建造了新的水污染控制工厂,以改善位于城市下游的部分湖泊的水质。大坝实际上是从 1979 年开始建造的,但是在苏联解体之后该工程被中断,直到 2011 年 8 月才完成。仅 4 个月后,屏障就帮助圣彼得堡免于遭受破坏性洪水的袭击。根据《莫斯科时报》的报道,自那时以来,屏障已成功地防止洪灾 13 次。自 1982 年以来,伦敦的泰晤士河屏障保护了伦敦市中心免受泰晤士河上涌的洪水袭击。自那时以来,为了防止洪水泛滥,人们已经又提高了屏障。

考虑到圣彼得堡和伦敦的屏障都是早就设计好了的,而当前关于气候变化的预测发生在这以后,因此要想继续发挥它们的作用,就必须对其进行实质性的重建,以提升高度。但是,这些例子表明了位于大型河口的城市在未来几年中将需要的各种保护措施。

根据气候科学家对随着海平面上升和风暴变得越来越频繁而激烈,而引发洪水的未来风险的评估,中国位于沿海和沿江的城市可能会受益于类似圣彼得堡或泰晤士河的屏障设施。

屏障的替代方案包括扩大土地面积或湿地以缓冲洪水潮,现有市区的防洪排涝系统可以起到保护作用。这些设施可以与其他基础设施的改造结合起来,以增强城市和区域的整体设计(图 5)。

按荷兰正在实施的设计政策,河流防洪应该"为河流腾出空间"。这些措施包括通过将堤坝移至离河流较远的地方来扩大洪泛区,重建防洪设施,拓宽河道,兴建新的运河以防止河水泛滥,以及重新设计河流构造(或桥墩等),以防水流出现倒流并蔓延。

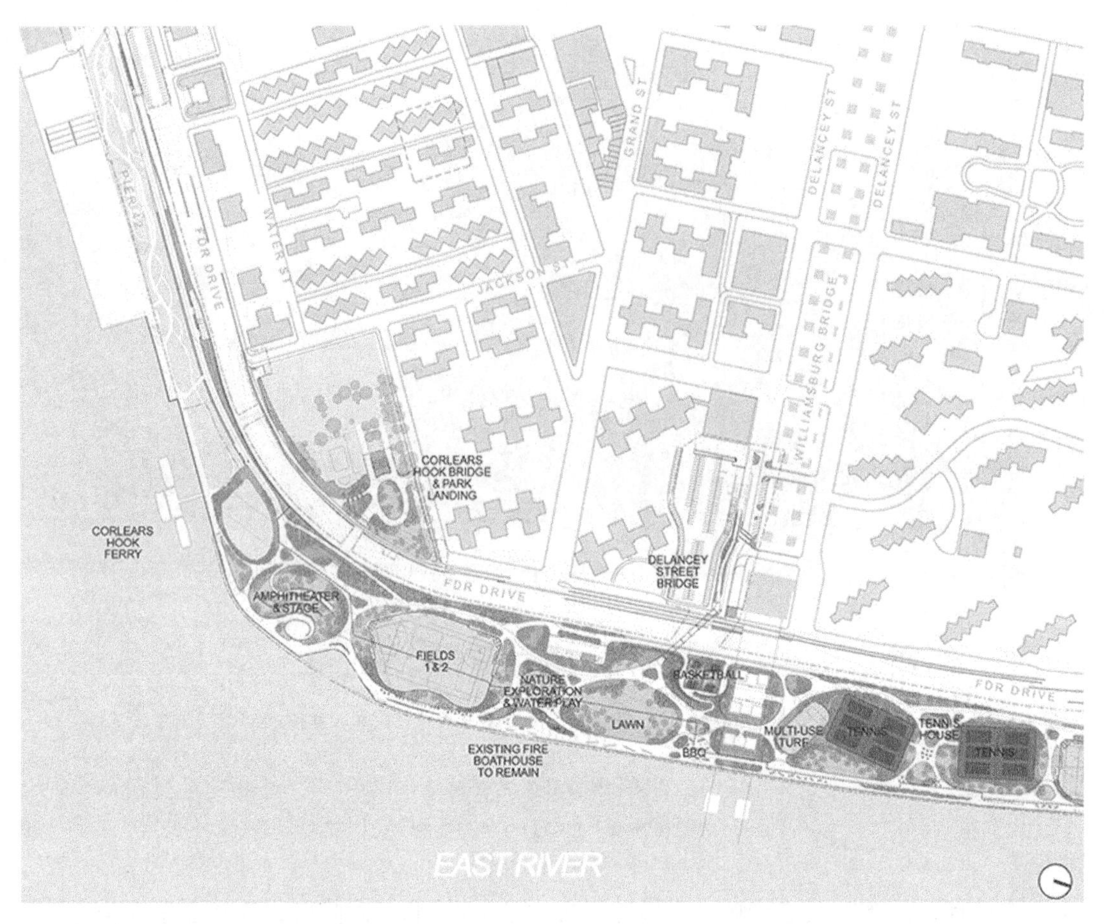

曼哈顿下城附近的东河正在修建防洪设施，该地区在2012年的一场暴风雨中遭受严重洪灾袭击。东河边缘有一个海堤，其后方的区域被修建成一个公园，人们在这里活动的同时，并不会意识到这里其实是防波堤。

图5　曼哈顿下城附近正在修建的防洪设施

Fig.5　Flood protection being built into the East River around lower Manhattan

应该设计和建造新的绿色基础设施，用于管理极端的降雨事件，这包括建设蓄水区和吸收区作为蓄水层，以及减少热岛等。这就需要绿色街道、绿色停车场和从废弃区域中开垦的公园。这些措施可以截留和容纳洪水，从而减轻河流洪水的影响，否则在洪水时期这些水会涌入河道(图6)。随着城市开放空间对于解决气候问题变得越来越迫切，所有社区都应该制定公共开放空间规划，将公园、街道和广场整合为一个统一的模式。设计指南应将建筑物与公共开放空间规划联系起来。

气候变暖还将导致更长更严重的干旱时期。应当在建筑物和农业中采取节水措施以防备干旱。一些沿海地区可能需要海水淡化厂，不过有必要开发比今天更好的盐水处置净化方式。与新加坡一样，废水的再利用可以作为增加饮用水供应的方式。

不建造海水淡化厂，而是在河流入海之前将水蓄起来，以确保淡水供应。这在某些地方可能是首选策略。也许可以将新加坡那样的蓄水坝与圣彼得堡那样的防洪堤结合起来(图7、图8)。

在汉堡易北河边的港城地区，人们设计了马可·波罗台地，考虑了公园靠近河的部分可能会被淹没的问题，如这张照片所示。当洪水消退时，公园很容易被清理出来，恢复为公共用途。

图 6　汉堡易北河边港城地区的马可·波罗台地

Fig.6　Marco Polo Terraces in Hafen City along the Elbe River in Hamburg

横跨新加坡滨海湾的水坝将海湾变成了淡水蓄水池，以增加城市的饮用水供应。大坝将海水挡在外面。多余的水会在退潮时被抽走。

图 7　横跨新加坡滨海湾的水坝

Fig.7　A dam across Marina Bay in Singapore

野火与上升的海平面不同，毕竟是有规模局限的。在森林或草原燃烧之后，可以发生新的生长，这将更为适合变化的气候，并且不太可能再次燃烧。在不久的将来，为了使开发区防火，有必要在附近规划荒地和设置禁建区。此外，在荒地附近建造的建筑物的设计应尽可能防火。在易遭受火灾的荒野地区内，不应继续建筑和占用。不过随着时间的推移，这些禁令也可以被取消，野火可能并不会像海平面上升和风暴潮那样导致大规模的人口迁移。

所谓的“雨水花园”被用于蓄水，直到风暴结束后再排出雨水并使其渗入地下。如果没有雨水花园的调蓄作用，直接将雨水通过管道系统排走的话，则有可能增加其他地方洪水。

图 8　纽约州雪城大学(Syracuse University)的绿色基础设施

Fig.8　Green infrastructure at Syracuse University in New York State

城市和农村的温室都应提高粮食产量并帮助防止粮食短缺。可以在具有必要支撑结构的建筑物(例如工厂和仓库)屋顶上建造城市温室。在农业地区增加温室可以提高生产力并使农作物多样化(图 9)。

在城市地区，有许多以这种方式种植食物的机会，这可以帮助在需要的地方提供食物，而无须支付大量的运输费用。

图 9　在加拿大蒙特利尔一个现有仓库的屋顶上建造的城市温室

Fig.9　An urban greenhouse built on the top of an existing warehouse in Montreal, Canada

4 有必要立即采取政府行动

所有这些措施的支出总和将会非常大，但是可以考虑分阶段实施。除了准确的成本估算之外，还需要有一个分阶段筹措和花费必要资金的计划，以便可以控制成本。

如本文所述的措施也需要纳入现有的开发法规中，在必要时需要编写新的法规。城市设计教育和实践也将需要发展以适应这些新情况。

气候变化不会停止。如果当前的趋势继续下去，到本世纪末变化的尺度将成倍增加。根据目前的预测，在本世纪中期的气候变化条件下安全地生活将是困难的，但仍然是有可能实现的。如果全球温室气体排放量不能（尽快）大幅度减少，那么在2100年的气候中人类安全地生活将是困难的，甚至根本是不可能的。

说明：

* 本文中图片均来自巴奈特教授在2019年10月29日在第一届"以人为本的城市设计"国际会议中的演讲稿。

参考文献

[1] 俞孔坚，李海龙，李迪华，等.国土尺度生态安全格局[J].生态学报，2009，29(10)：5163-5175.

Designing for people in a warming climate

Jonathan Barnett

People-oriented urban design policies are a correction for the years where China gave priority to efficiency and economy in an enormously successful process of economic and urban expansion. Now China has reached a point where the quality of the built environment claims a larger share of government attention. But this concern, while important, is looking backward to evaluating and improving what has already happened. Looking forward, major changes in the climate have already begun, and the consequences will become inescapable. Designing for people in a time of climate change includes making people's safety a very high priority as well.

There are seven major categories of threat from living in a warming planet: sea level rise and storm surges, river flooding, extreme rain events, life-threatening heat, drought, food shortages, and wildfire. Until recently, these problems were not expected to become significant until mid-century or even later. However, experts underestimated how quickly the climate would change. Recent newspaper headlines describe a seven-meter storm surge devastating the Bahama Islands in September 2019, only 320 kilometers from highly vulnerable Miami Beach, which is only 1.5 meters above sea level. It is becoming a question whether Venice can survive repeated tidal flooding after unusually high tides in 2019. Record river flooding has recently devastated many places, including the American Midwest in 2019. Torrential rains around Houston, Texas have caused one 500-year flood event and two 100-year flood events since 2016. June 2019 saw a record heat wave over much of Europe. Capetown, South Africa almost ran out of drinking water in 2018. Drought and food shortages are causing record movements of people away from rural areas where crops have failed, including people fleeing several countries in Central America—creating a refugee crisis for Mexico and the United States. Wildfire has devastated parts of California in 2018 and 2019, and is even taking place above the arctic circle in Sweden.

Figure 1 Professor YU Kong-Jian of Peking University and colleagues LI Hai-Long, LI Di-Hua, QIAO Qing, and XI Xue-Song published a study in 2009 entitled China National Ecological Security Patterns: A Framework for Reinventing the Good Earth. This study looked at preserving natural systems, the potential for ecological restoration, and the places where it is suitable to build for all of China.

Figure 2 Professor YU Kong-Jian of Peking University and colleagues divided the problem of preserving and restoring natural systems into five ecological security patterns: for headwater conservation, for flood control, for combating desertification, for soil and water conservation, and for biodiversity protection. Four or five subsystems within each category were mapped using a Geographic Information System program, and then put together by the program to create a map that optimized these factors for each security pattern. Then the five security patterns were put together again by the computer program, to produce an optimized map for ecological protection for the entire country. This pioneering work is a context and a model for looking at how to preserve natural systems and build in harmony with them.

The forces being unleashed by a changing climate should affect all urban design decisions about where to build, when to build, and what to build—not in 2050, or even 2030, but immediately.

There are already enough greenhouse gases in the atmosphere to require major changes in design and development practice during the next decade, regardless of efforts to limit future warming, and, by the middle of this century, a new practice of planning and urban design will have to be developed as a result of meeting climate challenges.

Securing people's safety will be essential to this transition process, producing many challenges for planners and urban designers.

1 Where to build safely

As much farmland as possible, not just prime farmland, should be preserved to increase food security. Regional design policies also need to be created to preserve and restore the natural landscape. The functional reasons for such policies will include retaining stormwater from rain events and keeping heat islands from forming, but landscape preservation and reconstruction are also needed to enhance the ability of all natural systems to withstand climate stresses.

New investments for infrastructure such as highways and railroads should take place in locations that will be safe from flooding or wildfire. Infrastructure located in places that are predicted to become unsafe need to be protected or rerouted. Regional design plans will be needed to prepare for these situations.

Extensions of existing communities, and the locations for new communities, should be constructed in places that are safe from flooding and wildfire, and do not detract from agriculture. Again regional designs and plans are needed to identify places where development should take place.

Existing coastal and riverfront cities, and their related infrastructure—including highways, railroads, and power plants—will need engineered protection from flood-risks to defend existing and new construction. Including the most appropriate urban design in the engineering for these installations will be important.

Places and infrastructure at risk from wildfire will need to be protected, and new development in such locations ought to be severely restricted, again as identified in regional designs.

Places where protection of existing buildings and infrastructure from flood or wildfire will not be feasible, today or by 2050, will need to be redesigned for phased withdrawal and clearance, beginning with rules that limit new construction, followed by reductions in government services, plus programs to help people and businesses to relocate. The decisions about how to help people deal with these major changes to their lives will be extremely difficult.

Alternate locations for populations that will need to move from endangered areas need to be identified in large-scale urban design plans, with priority given to improving by-passed and under-used areas, if they meet safety requirements, rather than developing green-field sites.

In both new and existing urbanized areas, land use and transportation should be integrated in order to reduce growth pressures on the natural environment and on agricultural land. The highest concentrations of commercial development should be directed to the areas around high-speed rail stations, and these stations should have good connections via transit to airports. Residential neighborhoods should be designed with access to transit. BRT lines can be developed in existing highway corridors in places where a fixed-rail investment will not be cost-effective.

2 When to build safety measures

Time to begin construction for engineered protections for vulnerable urban areas need to be scheduled and funded, considering that the necessary construction can take decades. Such protections need to be designed to meet the safety requirements of specific areas, as environmental conditions and predictions of climate changes vary significantly.

Timelines for withdrawal from endangered areas, beginning with a ban on new construction, also need to be scheduled, understanding that this process can also take decades, and that alternate locations should be as safe as possible from major climate threats.

A basic principle of environmental protection: new areas should not be urbanized until the full potential of existing urban areas has been realized. Recognizing this principle will mean major changes in the way urban growth is managed. New policies should set timelines for reinvestment in by-passed and under-used areas.

3 What to build

To reduce the emissions causing global warming, new urban investment should be in walkable, transit-rich, zero-carbon communities to reduce energy use, water use, and lower the need for automobiles. Urban designs should include managing recycling, and zones for solar and wind access. Well-known examples include the Hammarby Sjostad district in Stockholm and False Creek South in Vancouver, Canada. Designs for retrofitting existing communities to make them as energy efficient as possible will also be a priority. Both of these examples include a community-based small-scale power plant which makes use of heat transferred from hot water used in bathrooms, dishwashers, and washing machines, and from the wastewater pipes, to reduce the amount of fuel needed by the power plant. Biofuel for the local plant can also be created from the organic waste collected in the community.

Both large-scale and distributed energy alternatives should be put in place. Wind and solar farms should be built where the land is available, suitable, and most appropriate. Solar and wind arrays on individual buildings can reduce the energy losses from the transmission. Effective battery storage is now available so that renewable energy can remain available day and night.

For big coastal cities, flood surge protection can include engineering works like the St. Petersburg Barrier in Russia or the Thames Barrier in London. The St. Petersburg Barrier separates much of the estuary of the Neva River from the Gulf of Finland. The sequence of dams is interrupted by two movable gates—modeled on the gates that protect Rotterdam—to let shipping through and six sets of flood-surge barriers, which normally let the river water flow out, but can be shut when there is a flood warning. The top of the dam completes a ring road around the entire city, with tunnels under the shipping gates. New water-pollution control plants have been constructed within the barrier to improve water quality in what is now partially a lake, situated downstream from the city. The barrier was actually begun in 1979, but construction was interrupted after the dissolution of the Soviet Union, and was not completed until August 2011. Only four months later the barriers saved St. Petersburg from a damaging flood. According to the *Moscow Times*, the barriers have been used successfully to prevent flooding 13 times since then. London's Thames barrier has protected the center of London from flood surges coming up the river Thames since 1982. The barriers have been raised to prevent flood surges almost 200 times since then.

Figure 3 Plan of the barrier at St. Petersburg in Russia which can be closed to keep flood waters from the Baltic Sea from coming up the river Neva into the city. Normally the gates are kept open to let the river flow out and ships to enter and leave.

Figure 4　The flood barrier has been incorporated into St. Petersburg's ring road. The highway runs on the top of the barrier. At the gates the road goes into tunnels under the river.

Both the St. Petersburg and London barriers were designed long before current predictions about climate change, and both will need to be substantially rebuilt to make them higher, if they are to continue to be effective. However, they are demonstrations of the kinds of protections that will be needed by large, estuary cities in the coming years.

There are coastal estuary cities and coastal riverfront cities in China which may benefit from installations comparable to the St. Petersburg or Thames barriers, depending on estimates by climate scientists of future local dangers from flood surges as sea levels rise and storms become more frequent and more intense.

Alternatives to barriers include extensions to landmass or wetlands to absorb flood surges, and pump systems that can help protect existing urban areas. These installations can be combined with other infrastructure improvements to enhance the overall design of the city and region.

Figure 5　Flood protection is being built out into the East River around lower Manhattan, an area which was hit by severe flooding during a storm in 2012. There is a seawall at the edge of the river, and the area behind it is built up into a park, which people will use without ever thinking that it is also a flood wall.

River flood protection should "make room for the river", following design policies being implemented in the Netherlands. These measures include enlarging flood plains by moving levees and dykes farther from rivers, rebuilding flood protections, widening channels, providing new channels for times when rivers would otherwise flood, and redesigning areas where the configuration of the river—or such elements as bridge abutments—can obstruct the flow of water and cause it to back-up and spread out.

New green infrastructure should be designed and constructed to manage extreme rain events, including holding and absorption areas to recharge aquifers, as well as to reduce heat islands. Green streets, green parking lots, and parks reclaimed from disused areas will all be needed. These measures can intercept and hold water which would otherwise flow into rivers during times of flooding, and can thus lessen the impact of river floods. As urban open spaces become more necessary to deal with climate issues, all communities should have a public open-space plan that integrates parks, streets, and plazas into a single, coherent pattern. Design guidelines should relate buildings to the public open space plan.

Figure 6　At Hafen City along the Elbe River in Hamburg, the Marco Polo Terraces have been designed so that the parts of the park close to the river can flood, as seen in this photo. When the flood subsides, the park is easily cleaned and restored to public use.

A warming climate will also produce longer and more severe periods of drought. Water-saving measures in buildings and in agriculture should be put in place to prepare for drought. Some coastal areas may require desalination plants, although disposition of the brine that results from the purification process needs to be solved in better ways than are in use today. Re-use of wastewater, as in Singapore, can be ways to increase the supply of available drinking water.

Impounding a river before it enters the sea to secure a supply of fresh water, as opposed to building desalination plants, may be a preferred strategy in some locations. An impoundment dam like the one in Singapore could perhaps be combined with flood-surge protection as in St. Petersburg.

Figure 7　A dam across Marina Bay in Singapore has converted the bay into a fresh water reservoir to augment the city's drinking water supply. The dam keeps ocean water out. The excess river water is pumped out at low tide.

Figure 8　Green infrastructure at Syracuse University in New York State. What is called a rain garden holds water until the storm is over, allowing it to seep into the ground, rather than flowing away in a pipe system and possibly adding to flooding elsewhere.

Wildfire, unlike rising seas, is eventually self-limiting. After forests or grasslands burn, new growth can take place which is more suitable to the changed climate and less likely to burn. For the near future, cleared zones should be designed and implemented around wildland where needed to protect nearby development from wildfire, and buildings constructed near wildland areas should be designed to be as fire-resistant as possible. Building and occupation should not continue within wildland locations subject to fire. However, these prohibitions can be lifted over time. Wildfire will probably not require the large-scale movement of populations that can be necessitated by sea-level rise and storm surges.

Greenhouses, both urban and rural, should enhance food production and help guard against food shortages. Urban greenhouses can be built on existing rooftops of buildings which have the necessary support structure, such as factories and warehouses. More greenhouses in agricultural areas could enhance productivity and diversify crops.

Figure 9　An urban greenhouse was built on the top of an existing warehouse in Montreal, Canada. There are many opportunities to grow food in this way within urban areas, which can help to supply food close to where it is needed, without big transportation costs.

4　Government actions needed immediately

The totals for all these measures are going to be very large, but the expenditures can be staged. What will be needed, in addition to accurate cost estimates, will be a plan for raising and expending the necessary funds in stages, so that costs can be manageable.

Measures like those described in this article also need to be incorporated in existing development regulations, and where necessary, new regulations need to be written. Urban design education and practice will also need to evolve to meet these new situations.

Climate changes will not stop. If current trends continue, changes by the end of the century will be exponentially greater. Living safely with the climate changes already predicted for mid-century will be difficult but possible. If global greenhouse gas emissions are not reduced drastically—soon—living safely with the climate in 2100 will be difficult and may not be possible at all.

德国的以人民为本城市更新：
从实体提升到本地的共同生产

People-oriented urban regeneration in Germany:
From physical upgrading to local co-production

［德国］魏尔·阿尔陶克
Uwe Altrock

本文介绍了德国依靠公共资金扶持的城市更新传统及其对以人民为本的城市发展所做的贡献，这涉及对待现有邻里、城市肌理以及内部社区的处理方式的重大转变。文章首先着眼于更新的方法，然后会总结那些非常复杂的微观干预措施，它们是在规划师、当地利益相关者和居民共同作用下形成的。文章将讨论为什么当前的方法更侧重从社会经济方面稳定社区，其方法的主要策略是什么，以及它对克服城市更新的挑战有何贡献。本文的目的是说明以人民为本如何成为成熟城市中城市更新的关键要素。这非常适合解决城市内部人口的需求，特别是那些面临可持续性挑战的受忽视地区。文章基于德国长期以来对城市更新实践的研究，部分资料来自2017年完成的针对"'社会融合城市'城市重建资金计划"的中期评估，该项目受德国联邦政府委托(BBSR & BMUB, 2017)。

在这个背景下，城市更新被理解为"全面和整合起来的愿景和行动，致力于解决城市问题，并力求使目标地区的经济、实体空间、社会和环境状况得到持久的改善"(Roberts, 2000:17)。城市更新已经发展成为一个重要的政策领域，拥有精心制定的法律框架和相当稳定的公共资金支持，在与中国情况比较时，必须注意这个前提条件。第一，城市的重大危机会严重(且有选择地)影响内城地区的邻里。当地的许多人遇到各种困难，比如曾在战争中遭受摧残、受到意识形态方面的排斥(例如存在偏爱新建筑的政治倾向，不愿意维护现有的住房，因而在管理和维护老旧建筑物方面存在法律障碍)、某些区位因素的变化，或者由于城市改造而导致(私有)房地产所有者认为经济前景不佳，而不愿意进一步向建筑物维护投资。第二，政治和经济体系允许政府对邻里发展进行实质性干预，并允许引入公共资金来改善一个地区，从而恢复私人投资者的信心以撬动私人投资。该机制为德国的长期区域重建提供了资金，并在几十年里获得了稳定的政治合法性(BMVBS, 2011; Busch et al, 2018)。第三，尽管在第二次世界大战后，德国经历了经济快速增长，先前的城市发展也积累了大量的资产，但是仅仅依靠市场力量和潜在的利益取向已经不足以维持市中心地区的活力，因此有充分的必要针对现有的城市地区制定专门的城市政策。

1 背景：德国城市更新的悠久历史

德国的城市更新始于1960年代初期，这是对第二次世界大战中内城遭到破坏以及对那些建于1900年快速工业城市化时代的老城和邻里的价值忽视的反思。由于这些地区当时的生活条件恶劣，结果遭受污名化，战后时期的现代主义建筑和城市设计运动趁机与半国有的房地产公司和地方政客结成联盟，以清除那些1900年左右所谓的不合格密集租屋住宅。

战后时代，德国城市可怕的住房状况得以改善，而这主要是通过在城市外围建造新的大型居住区，但很快又涉及城市更新方面的任务：在市中心条件恶劣的租屋公寓楼中的居民获得承诺——为他们在市郊提供更好的住房，因此同意拆除当时被认为过时的租屋公寓，结果城市内部大量租屋公寓被现代主义的新建筑所取代。

由此造成的流离失所和对邻里社会纽带的破坏，使得这项政策很快就遭到专家和市民的严厉批评。尽管如此，这种做法持续到1970年代初，当时德国已经正式建立了新的城市再开发体系，用于对划定的城市内部地区进行更新，整个城市开发补助基金由联邦和州政府共同出资。但是当时的法律框架和资金计划仍在支持拆除历史建筑群的行为，目标是为传统肌理的"现代化"创造条件。人们认为这些传统肌理由于地块面积小、功能和建筑物混杂、建筑密度高，再加上基础设施和服务设施的标准太低，常常会阻碍再开发的升级工作。

受到各地具体情况的影响，大拆大建的更新工作取得的成效有所不同。在一些城市中，整个邻里都被拆除，而在其他地方主要是历史中心的关键部分。由于内城区的居民主要是租客，因此他们在自上而下的规划中被理性主义者当作是被动的对象。结果是矛盾的：虽然他们的生活条件确实得到了改善，但其社会网络却遭到了破坏。由于补贴公共住房计划的实施，租户们要么会面临新公寓租金水平的急剧上升，要么是更新过程吞噬了大量的公共基金。此外，现代主义大型居住区内部缺乏宜人的氛围和有活力的公共空间。从长远来看，大型居住区会鼓励人们从以公交为导向变得更加依赖汽车出行。

2 以人民为本的城市发展，阶段1：谨慎的城市更新和国际建筑展

虽然这种发展对较小的城镇的影响不大，但它破坏了许多较大城市的宜居性。所获得的好评使得1970年代城市更新政策逐渐发生变化。由于当时学生政治运动的不断高涨，再加上石油危机的影响，以及1975年欧洲遗产年前后人们对历史建筑群越来越重视，德国整体的政治气候最终得到了改变。

柏林国际建筑展（IBA）对城市更新的影响达到了高潮，这是此类建筑展第一次把焦点集中在一个在二战中遭到破坏并在此后遭到忽视的内城地区。1960年代和1970年代人们开始对内城地区进行改造，但却没有能够产生令人满意的城市设计替代方案。在这种情况下，变革的主要动力来自居民和当地店主的活动，他们抵制大拆大建的简单方法。柏林的城市更新依靠居民咨询委员会的非正式支持，以制定和实施更新策略。这是使本地行动者加入更新过程

的第一个重要步骤,并由此增加了参与性要素。特别是,国际建筑展(IBA)在提出的愿景中确立了著名的"十二项原则"(SenBauWohn, 1991: 202f; Bodenschatz, 1987)。这些原则主张要求现有的居民和企业家一起制订更新计划,强调要尊重他们的"有形权利",因此要求更新要逐步进行,一点点完成。对于实体空间来说,应该把拆除限制在尽可能低的水平,同时通过在庭院绿化和外立面的装饰来改善现实状况,街道、广场和公园等公共设施应根据公共需求得到更新和保护。

这套方法于 1980 年代后期开始施行,并于 1990 年统一后推广到东德,现在已成为指导整个德国谨慎的城市更新工作的主要指南。大拆大建不再是基于城市更新的一部分。居民的权利以多种方式受到尊重,这里不仅是对土地所有者,而且还明确要求照顾到作为内城地区租户的权利,他们占了当地居民的大多数:保持当地建筑环境的做法得到认可,保护其社会纽带也被认为是值得的,并且理解当地居民没有能力负担利润最大化的升级措施。尽管从长远来看,环境提升仍然会逐步导致居民的外迁,但参与性元素被系统地纳入了更新规划的过程中,并引入了一系列制度性要素,诸如成立居民委员会、向受影响的租户提供咨询服务等,这有助于重点关注内城地区居民的需求。不过更新还是把重点放在实体环境的升级,旨在恢复被忽视地区的投资环境并解决城市问题。因此可以总结出,"谨慎的城市更新"这一构想是在公共资金的帮助下使城市更新地区"恢复正常"。

3 以人民为本的城市发展,阶段 2:社会导向的更新和"社会融合城市"计划

在整个 1990 年代,人们认识到,德国各州之前单纯通过升级实体空间的政策,已经无法再稳定那些遭受贫困和失业困扰的邻里。一些有"城市问题"的地区刚刚得到升级,但是并没有表现出社会经济复苏的迹象,那些较富裕的居民仍然在选择外迁,这使得社会处境不妙人口的集中问题变得更加严重。为此人们提出了一种新的升级方法,要求把重点放在社交网络、赋权、教育、社会与种族融合以及创造就业方面。其背后的主要思想在于:社会经济剥夺有空间维度,也就是说,许多消极指标在空间上会显著集中,贫困程度高的弱势社区遭受了负面的污名化。因此,在旨在改善社区状况的公共城市更新资金的帮助下,似乎有望解决社会空间剥夺问题。但是,这些资金属于联邦和州城市开发部预算的一部分,因此被视为投资类项目;相反,社会项目(培训、资格认证、社会工作等)则是消费性的项目,属于地方当局或者由联邦和联邦州负责社会事务部门的职责,而且通常的做法是把经费发给个人,而非拨给邻里。

这导致了 1999 年"社会融合城市"计划的出台。该计划是希望通过以下五种方式将实体空间升级与社会经济升级相互结合起来:①该程序旨在识别和升级关键建筑物,并采取适应性方式对其进行再利用,这些设施可以最好地满足当地社区的需求,并促进社会互动。计划坚持认为,通过投资某些实体空间结构,可以促进各种社区活动、居民互助和参与,这就可以支持公共生活的发展。②通过将其他旨在促进社会生活的公共资金来源用于施行社会融合城市计划中的邻里,人们希望采用整合性的方法,共同努力改善邻里,支持社会弱势群体,并进行实体空间升级。③将实体升级方面和社会措施结合起来,制定整合性的邻里发展规划,以确保来自不

同计划中的公共资金共同促进整体的空间发展战略，避免相互矛盾的活动，并最大限度地减少不同职能部门之间的冲突。④安排邻里管理团队负责协调各种利益相关者、融资计划和活动，该团队还应帮助倡导人们的利益，在不同利益相关者之间建立网络，调动更多资金，组织各种参与和授权活动，从而改善邻里形象。⑤为了加强对参与和赋权的充分关注，将少部分可获得的资金分配给所谓的“邻里基金”，资金会用于那些由当地居民和店主自主提出的项目建议，经由当地委员会选拔以后，这些项目就可以实施，而无须再受地方政府或其他资助机构的约束。

除了这个战略性程序框架外，还有许多建设任务模块也属于社区融合城市计划的行动领域的一部分(BBSR & BMUB, 2017:37ff)。比如传统的实体空间升级措施，侧重于改善住房条件及其周边的公共和私人空间质量，这是因为废弃的建筑物往往会是造成邻里名声不佳的重要原因。以人民为本的特征体现诸如基于需求对公寓建筑物内部及其建筑群的布局进行调整、建设跨文化交流的花园，其作用是通过重组公共空间来提供多功能设施，并促进居民的社会融合(这方面的目标很难通过直接干预来实现)。

资金可以帮助解决许多问题，包括改善地区的宜居性，扩大当地居民的发展机会等。这首先就涉及社会与种族融合的问题。为此，计划特别考虑到改善教育设施和社会基础设施等内容，而幼儿的教育和社会弱势群体的职业培训等问题对确保机会均等和少数族群的融合至关重要。在许多情况下，城市规划与学校部门之间会密切合作，旨在把实体空间措施和社会措施结合起来，使学校成为名副其实的邻里中心。为此，人们会对学校建筑进行改造，将其与学龄前服务设施一起考虑，并向社区开放。雄心勃勃的城市政府建立起复杂的教育网络并改革了教育体系，以更好地支持弱势群体家庭或少数族裔的孩子。

为了支持社会与种族的融合，人们还大力支持发展多元文化中心和青年中心，提供自助和互助导向的设施，并举办街头节日活动。由于受忽视地区经常遭受经济转型和失业的打击，该计划还试图促进当地(通常与种族有关)经济。为此，本地企业与教育和培训机构之间在职业资格方面的合作关系就非常重要。对新企业和种族相关经济的扶持主要是通过辅导和培训计划实现的。在城市营销活动方面，重点是提升体现民族特色购物街的形象。除了这些措施之外，雄心勃勃的邻里还会制定包括交通、环境、文化、体育、公共卫生与安全领域的战略。公共关系策略还致力于克服该地区的污名化并提升形象。

4 社会融合城市计划的成就和不足

德国各地已经有 900 多个指定区域获得了公共资助。如果要评估其内容，则必须注意资金应用的多样性，地方政府可以分别根据自己面临的挑战来制定更新策略，并确定目标。上文提到的行动领域可以被看作是一系列资助计划的子策略集，只要当地有需要的话，就可以选用这些子策略。虽然在许多更新地区，住房和公共空间都是主要内容，但在社会与种族融合领域，人们可以发现社会融合城市计划的重大创新。计划评估已经证实，资金确实加强了社会网络和各种动员举措(BBSR & BMUB, 2017)。

社会融合城市计划围绕以人民为本的制度安排进行了深入探索，有助于加强赋权，并大大提高了升级项目的参与性。其中主要原因包括：邻里管理强调赋权导向，社会基础设施的催化作用以及邻里基金为各种微型项目提供机会。但是,这些影响还不能完全克服城市规划中参与的选择性问题:由于在社会资本和时间等方面的资源差异,参与活动的人群往往只是一部分居民,不仅一般的规划参与中是如此,而且社会融合城市计划也遇到这方面问题(Selle, 2013; BBSR & BMUB, 2017)。

在评估机构创新的作用时,有明确的证据表明,这些创新已得到当地利益相关者的成功实施和赞赏。在大多数情况下,整合发展构想在程序方面的安排能够使各种地方政府的机构和市民参与进来:例如在91%的案例中,社会事务部参与了由规划部门指导的整合发展构想的准备工作,相比之下,公园管委会和学校部门的参与率为79%,儿童和青少年福利服务部门是78%,交通管理部门为72%(Difu, 2015; BBSR & BMUB, 2017: 48)。在所有案例中,地方政府的组织参与率是88%,学校参与率为82%,住房公司参与率为78%(Difu, 2015; BBSR & BMUB, 2017: 49)。但是,体现以人民为本的关键,是在所有案例中有85%成立了邻里管理办公室。尽管他们在更新工作的管理方面有许多不同的任务,但他们的工作重点是赋权:大约95%的社区管理人员正在积极贯彻赋权,并参与当地居民的活动,向公众发布有关更新工作的信息,充当居民的一站式服务办公室并协调邻里行动(Difu, 2015; BBSR & BMUB, 2017: 55)。

总而言之,新计划以五种不同但相互关联的方式来处理城市问题,并更加关注社会方面:①它以相当传统的方式通过提高实体空间环境的标准和状况来改善当地的生活条件。②它借助更好的(社会)基础设施和服务设施,改善了当地居民的生活,增加了发展机会。尽管城市更新的早期阶段已经朝着相似的方向努力,但社会融合城市计划将其大部分精力集中在这一领域,并动员了更多的公共资金用于这些活动。③它通过积极的榜样和社会学习机会,提高了社会融合的可能性,这一点尤为重要,因为社会处境不利地区经常会出现大量辍学,而青少年由于本身缺乏职业方向而错过职业选择的机会,最终可能会失去社会融合的条件。④可以经常通过改善邻里的形象,来避免或减少将来的污名化和歧视。⑤通过加强互助组织的能力,并提高人们对社会弱势群体的政治关注度,可以扩大被忽视地区的影响力,并有可能在地方政府层面的政治决策中更好地考虑到他们的需求。

尽管通常来说,社会融合城市计划中的城市更新可以有效地恢复人们对受资助地区未来前景的信心,但由于具体项目和措施仍然有可能受到财务和制度方面的限制而具有某些不确定性:城市更新基金旨在帮助指定区域改善和稳定局势,可以保持10年甚至15年的资助跨度。不过在那之后,城市更新地区应该实现蓬勃发展,而无需额外的资金。考虑到在实施社会融合计划的城区内部,往往需要长期的稳定和支持措施来加强社会网络,因此很难达到上述期望。相比之下,升级实体空间结构的措施(邻里中心、体育设施等)通常确实对社会与种族融合产生积极影响。当人们审视经济影响时,就不得不承认,稳定需要漫长的耐心,而在这种情况下成功的案例很少。值得注意的例子主要是民族购物街的正面形象和青少年的培训。在经济较弱城镇内部,邻里面临居民有选择外迁的情况并没有被真正克服,其中主要的驱动因素是不稳定和污名化问题。不过最近"再城市化"的大趋势已经在许多城市表现出来,这说明内城邻里对雅皮士甚至普通家庭又变得非常有吸引力。在这种情况下,可以注意到,社会融合城市计划推动了许多邻里重新获得吸引力。另外,那些并不富裕的城市仍然面临严重的社会空间两极分化,当地甚至存在贫困孤岛。尽管诸如社会融合城市等计划具有稳定作用,由于针对特定地区的更新资金有限,因此很难应对社会向后工业时代转型面临的普遍挑战。

5 结论：城市更新中的共同生产

通过历史回顾可以看到，以前自上而下的规划模式正在不断转型，致力于使居民能够更好地融入更新战略的准备和实施工作。早期的经验表明，大拆大建的更新往往导致缺乏活力的结果，兴建的现代主义居住区缺乏吸引力且毫无特色。如果用精英的豪华公寓取代传统社区，要么是出现不尊重社会纽带的问题，要么就是利用社会弱势群体财力有限的缺陷。从1960年代后期开始，更多的整合性和参与式方法表明，可以采取更加谨慎的方法进行更新工作，它尊重当地居民的需求并争取邻里具有宜居性。逐步使历史街区适应当代需求的微更新方法，不仅要依靠当地知识，还要建立在对城市内部生活充分调查的基础上。这些努力带来了制度上的创新，帮助改变了自我观念，并提升了参与性规划的重要性。在分析了城市更新可能的负面影响之后，人们又引入了居民委员会的制度。制定更新策略时必须听取居民委员会的意见。随着干预的重点从实体空间升级转向更加整合性的方法，包括应对社会经济剥夺、提升社会弱势群体的机会、通过邻里管理人员积极促进赋权等，此外还包括扶持由当地居民管理和实施的微型项目。

自20世纪末以来，这种大趋势已嵌入德国等其他国家城市规划的重大转变中。研究合作性规划的理论家，例如帕齐·希利(Patsy Healey, 1997)或德国的克劳斯·塞拉(Klaus Selle, 1994, 2005)已经阐述了这种趋势。在过去的20年中，已经产生了许多新的创新性规划方法，这些方法可以归入"共同生产"概念中。这个认识清楚地表明，城市空间不再仅仅被看作是实体空间的，同时也是社会性的建构过程；人类的互动仍然要发生在欧几里得式的三维空间当中。这种认识使得最近在"城市权"大旗下发展的地方倡议合法化(Lefebvre, 1968)，声称城市的规划和空间利用不应局限于较低的参与层次(Arnstein, 1969)，人们有权"共同创造"空间。这些举措在内城蓬勃发展，这是因为内城地区的空间有限，空间的使用常常受到所有权和私人利益的限制。尽管社会融合城市计划通常仅限于城市中的小片区域或仅仅一个地块上面，但它们代表着对城市更新的不同理解，在这里充分赋权的人们可以发展出替代性的城市生活形式。在后工业转型中，它们有时不得不被视为向停滞地区城市内部废弃地区注入新生命的唯一手段。

这些举措有多种形式。一种主要的方法是依靠小规模的互助社区团体进行自下而上的活动，以保存那些空置但具有历史意义的重要建筑物，例如德绍的所谓"施瓦本住宅(Schwabehaus)"(www.schwabehaus.de)或位于格赖夫斯瓦尔德(Greifswald)的"Straze"项目(www.straze.de)，这两个不断发展的非营利文化中心都向当地公众开放，其中多亏大量个人奉献精神的支持。另一个方法是在萎缩的城市中对闲置的空地进行改造和引入临时功能，以用于公园、城市园艺或其他活动，它们的内涵超越了利润导向的城市生活(Oswalt et al, 2013; BBR, 2004)。最近，这种方法得到了蓬勃发展，但是当利润回升、私人土地所有者的利益又重新被提及以后，就开始与常规的规划内容出现抵触(www.prinzessinnengarten.net)。城市干预表明了内城地区如何实现新的生活，有时甚至会在规划师、建筑师和当地居民之间启动新的联盟(Altrock & Huning, 2015)。

重要的问题在于，人们在50年前建立了城市发展扶持基金框架，其核心特征是在成熟城市中帮助不断衰落的邻里得以升级和稳定。对于这些邻里来说，即使在与当地合作生产的情况下，私人开发商也很难取得收益。在基金框架的支持下，各种新型和试验性的城市生活形式得以发展，这既包括在社会融合型城市推广赋权的措施，也包括自下而上的共同生产的努力。因此，除了在以利润为导向的中心商务区开发以外，城市更新及其公共资金体系会在内城地区坚持存量发展的背景下继续发挥至关重要的作用。

参考文献

[1] Altrock U, Huning S, 2015. Cultural interventions in urban public spaces and performative planning: Insights from shrinking cities in Eastern Germany[M]//Tornaghi C, Knierbein S. Public space and relational perspectives: New challenges for architecture and planning. London: Routledge: 148-166.

[2] Arnstein S, 1969. A ladder of citizen participation[J]. Journal of the American Institute of Planners, 35(4): 216-224.

[3] BBR, 2004. Zwischennutzung und neue Freiflächen: Städtische Lebensräume der Zukunft[M]. Bonn: BBR.

[4] Bundesinstitut für Bau Stadt-und Raumforschung (BBSR) im Bundesamt für Bauwesen und Raumordnung (BBR), Bundes Ministerium für Umwelt, Naturschutz, Bau und Reaktorsicherheit (BMUB), 2017. Zwischenevaluierung des Städtebauförderungsprogramms Soziale Stadt[M]. Bonn: BBSR.

[5] Bundesministerium für Verkehr, Bau und Stadtentwicklung (BMVBS), 2011. Wachstums-und Beschäftigungswirkungen des Investitionspaktes im Vergleich zur Städtebauförderung[M]. Berlin: BMVBS-Online-Publikation.

[6] Bodenschatz H, 1987. Platz frei für das neue Berlin! Geschichte der Stadterneuerung in der "größten Mietskasernenstadt der Welt" seit 1871[M]. Berlin: Transit.

[7] Busch R, Heinze M, et al, 2018. Perspektiven quantitativer Wirkungsanalysen. Ökonomische Effekte der Städtebauförderung: Quantitative Wirkungsanalysen in der Städtebauförderung[M]// Altrock U, et al. Stadterneuerung im vereinten Deutschland: Rück-und Ausblicke. Jahrbuch Stadterneuerung 2017. Wiesbaden: Springer: 275-294.

[8] Deutsches Institut für Urbanistik (Difu), 2015. Bundesweite Kommunalbefragung zur Sozialen Stadt [R]. Unpublished.

[9] Healey P, 1997. Collaborative planning: Shaping places in fragmented societies[M]. Houndsmills, England: Macmillian Press.

[10] Lefebvre H, 1968. Le Droit à la ville[M]. Paris: Anthropos.

[11] Oswalt P, Overmeyer K, Misselwitz P, 2013. Urban Catalyst: Mit Zwischennutzungen Stadt entwickeln[M]. Berlin: Dom Publishers.

[12] Roberts P, 2000. The evolution, definition and purpose of urban regeneration[M]//Roberts P, Sykes H. Urban Regeneration: A Handbook[M]. London: SAGE: 9-36.

[13] Selle K, 2013. Über Bürgerbeteiligung hinaus: Stadtentwicklung als Gemeinschaftsaufgabe? Analysen und Konzepte[M]. Detmold: Rohn.

[14] Selle K, 2005. Planen. Steuern. Entwickeln. über den Beitrag öffentlicher Akteure zur Entwicklung von Stadt und Land [M]. Dortmund: Rohn.

[15] Selle K, 1996. Was ist bloß mit der Planung los? Erkundungen auf dem Weg zum Kooperativen Han [M]. Dortmund: Dortmunder Beiträge zur Raumplanung.

[16] Senatsverwaltung für Bau-und Wohnungswesen Berlin, 1991. Internationale Bauausstellung Berlin 1987. Projektübersicht, Aktualisierte und erweiterte Ausgabe[M]. Berlin: SenBauWohn.

People-oriented urban regeneration in Germany: From physical upgrading to local co-production

Uwe Altrock

This paper deals with the tradition of publicly funded urban regeneration in Germany and its contribution to people-oriented urban development. It will talk about a major shift in the way existing neighborhoods, their urban fabric and the communities living in them were and are being treated, starting with approaches focusing on renewal but having arrived at very sophisticated micro interventions co-developed by planners, local stakeholders and residents. It will discuss why the contemporary approach rather aims at stabilizing communities socio-economically, what the major strategies of this approach are and what it can contribute to mastering the challenges of urban regeneration. The aim of the paper is to show how people-orientation became a crucial element of urban regeneration in maturing cities that seems appropriate in dealing with the needs of inner-city populations especially in disadvantaged areas confronted with the challenges of sustainability. For this purpose, the author builds on long-standing accompanying research of urban regeneration practices in Germany and an interim evaluation of the urban redevelopment funding program of "socially integrative cities" commissioned by the federal government and having been completed in 2017 (BBSR & BMUB 2017).

Urban regeneration in this context is understood as "comprehensive and integrated vision and action which leads to the resolution of urban problems and which seeks to bring about a lasting improvement in the economic, physical, social and environmental condition of an area that has been subject to change" (Roberts 2000: 17). That it has become an important policy field at all and that it can resort on both an elaborated legal framework and rather stable public funding has a number of prerequisites one has to keep in mind when comparing the situation for instance with China. First, cities have undergone major crises that have seriously (and selectively!) affected inner-city neighborhoods. Many of them have suffered from destruction in wars and/or neglect due to various reasons such as ideology (political preference for new construction instead of care for the existing housing stock, legal barriers for decent ways to manage and maintain aging buildings), changing location factors of certain areas or a negative economic outlook of (private) real estate owners due to urban transformations that have no longer let it seem worthwhile to take care of their buildings. Second, the political and economic system has allowed for substantial state intervention into neighborhood development and the invention of public funding to improve an area, thereby restoring private faith and thereby leveraging private investment, the mechanism has yielded public funding of long-term area-based regeneration , and a stable political legitimacy over decades in Germany (BMVBS 2011; Busch et al 2018). Third, despite the rapid economic growth in the era after World War Ⅱ in Germany, previous urban development had produced a number of economic assets that urban policies focusing on the existing parts of cities had become indispensable, as relying on market forces and potential profit orientation in real estate proved insufficient to sustain the vitality of inner-city areas.

1 Background: A long history of urban regeneration in Germany

Urban regeneration in Germany started systematically in the early 1960s as a reflection on the destruction of inner cities in World War Ⅱ and the neglect of historic old towns and the neighborhoods that had been built in the era of rapid industrial urbanization around 1900. The latter was negatively stigmatized due to the bad living conditions that had characterized them in their early years. The modernist movement in architecture and urbanism took its chance to coin an alliance with parastatal housing companies and local politicians in the post-war era to overcome the dense agglomerations of substandard tenement houses.

Major attempts to build new mass housing districts in the urban periphery served to improve the dire housing situation in German cities in the post-war era but were soon linked to urban renewal: Residents in substandard inner-city tenement buildings were promised to receive better new apartments in the periphery, thereby allowing for a demolition of the tenement buildings now considered obsolete to make way for replacing them by modernist new structures even in inner-city areas.

The resulting displacement and the destruction of social ties in the neighborhoods were soon heavily criticized by both experts and citizens. Nevertheless, when the official system of area-based urban redevelopment based on urban development grants for cities co-funded by the federal and state governments was officially established in the early 1970s, both the legal framework and the funding schemes still acted in favor of demolishing historic structures. They aimed at making way for a "modernization" of the traditional fabric that often seemed to stand in the way of upgrading efforts due to the small size of plots, a complicated mix of functions and buildings, and the low standard of infrastructure and amenities coupled with high building densities.

How far outright renewal efforts went depended on the local conditions. In some cities, entire neighborhoods were demolished, in others, key parts of the historic centers. As a major part of the residential population in inner cities were tenants, they were considered as objects of rationalist top-down planning. The results were ambivalent: While the living conditions were indeed often improved, social networks were destroyed and as a result of the application of subsidized public housing schemes, tenants either faced sharply increased rent levels in new apartments or the renewal process devoured lots of public funds. Besides, the modernist large housing estates lacked ambience and vibrant public space. In the long run, they supported the transition from a transit-oriented city towards a car-dependent city.

2 People-oriented urban development, stage 1: Careful urban regeneration and the International Building Exhibition

While this development did not take off as seriously in smaller towns, it affected the livability of many bigger cities. The critical acclaim it received gradually led to a change in the regeneration policies in the 1970s, when the upcoming politicized student movement in conjunction with the consequences of the oil crisis and the increasing appreciation of historic structures around the 1975 European Architectural Heritage Year had changed the general political climate in Germany.

Its impact on urban regeneration culminated in the International Building Exhibition (IBA) in Berlin, the first major exhibition of its kind focusing on an inner-city district that had been damaged by World War Ⅱ and neglected thereafter, an area where the first renewal efforts of the 1960s and 1970s had found one of its major stages yet without being able to produce satisfactory urban design alternatives. As a major driver for change in this context could be found in the local activities by residents and shopkeepers that resisted simple approaches focusing on demolition and renewal, urban regeneration in Berlin relied on resident advisory councils informally backing the development and implementation of regeneration strategies. This was the first major step towards involving local actors in the regeneration process and thereby adding a participatory element to it. In particular, the IBA promoted a vision that went much further when it formulated its well-known "twelve principles" (SenBauWohn 1991: 202f; Bodenschatz 1987). Those principles advocated for a regeneration that ought to be planned with the current residents and entrepreneurs, respecting their "tangible rights". Regeneration was to take place gradually and to be completed little by little. The physical situation ought to be improved by minimum demolition, by green development in courtyards, and by the decoration of facades. Public facilities such as streets, squares and parks should be renewed and preserved in accordance with public needs.

This approach was implemented in the second half of the 1980s and transferred to East Germany after the reunification in 1990. It became a major guideline directing careful regeneration processes in all of Germany. Major demolitions were no longer part of area-based regeneration. Rights of residents—that is, not only land owners but explicitly also tenants that made for the majority of residents in inner cities—were respected in various ways: by accepting that their built environment should not be destroyed, their social ties were worth being protected and that they did not have the means to afford profit-maximizing upgrading measures. Although upgrading nevertheless led to a gradual displacement in the long run, participatory elements were systematically included in the planning process of regeneration, and institutional elements such as resident councils and consulting services for affected tenants were introduced. This helped to bring in a focus on the needs of people in inner-city neighborhoods. However, the major focus of regeneration was a physical upgrading that aimed at restoring the investment climate in neglected areas and at overcoming urban problems. One can summarize that this idea is based on the will to "normalize" regeneration areas with the help of public funding.

3 People-oriented urban development, stage 2: The shift towards socially oriented regeneration and the program of "socially integrative cities"

Throughout the 1990s, policies for the physical upgrading in various German states locally hit by poverty and unemployment were no longer considered sufficient for the stabilization of disadvantaged neighborhoods. Some of the areas with a concentration of "urban problems" that had just been upgraded showed no sign of socioeconomic recovery and witnessed a selective outmigration of the more affluent parts of the resident population, making the concentration of socially disadvantaged become even more serious. A new approach towards upgrading with a stronger focus on social networks, empowerment, education, social and ethnic integration and job creation was invented. One of the major ideas behind it was that socio-economic deprivation had a spatial dimension, i. e. there was a noticeable spatial concentration of a number of negative indicators and that disadvantaged neighborhoods with a lot of deprivation were negatively stigmatized. Therefore, it seemed promising to address socio-spatial deprivation with the help of public urban regeneration funds directed at the upgrading of neighborhoods. However, those funds are technically part of the budgets of federal and state urban development ministries and therefore considered as investments. Social projects (training, qualification, social work, etc.) are, in contrast, seen as consumptive and therefore in the domain of municipalities or ministries that are responsible for social affairs but normally address individuals and do not spend their funds with a neighborhood focus.

This led to the newly-designed program of "socially integrative cities" invented in 1999. Its idea was to combine physical upgrading with socio-economic upgrading in five ways: First, the program aimed at the identification and upgrading and adaptively reusing key buildings in a way that they could best serve the needs of the local communities and as enablers for social interaction. The idea was that public life should unfold by investing in those physical structures that allow for a variety of community-based activities, self-help and participation. Second, by committing sources of other public funding aiming at the promotion of social life to directing their funds into the neighborhoods in the socially integrative cities program, one hoped for concerted efforts to supporting the socially deprived population together with a physical upgrading and thereby an integrated approach to improving neighborhoods. Third, this combination of physical and social measures was to be embedded into an integrated development plan for the neighborhood making sure that public funding in different schemes contributed to an overall spatial development strategy, avoiding contradictory activities and minimizing conflicts between different functions. Fourth, the complex task of coordinating the variety of stakeholders, funding programs and activities was laid in the hands of a team of neighborhood managers, present in the area to advocate for the people's interests, to build networks between different stakeholders, to mobilize additional funds, to organize participation and empowerment and to improve the image of the neighborhood. Fifth, the outright focus on participation and empowerment was strengthened by allocating a smaller percentage of the foreseen funding to a so-called "neighborhood fund" reserved for projects directly generated by the local residents and shopkeepers and selected autonomously by a local jury without involving the municipality or other funding agencies.

Besides this strategic and procedural framework, there is a number of building blocks that constitute the fields of action in neighborhoods that are part of the socially integrative cities program (BBSR & BMUB 2017: 37ff). The traditional repertoire of physical upgrading focusing on an improvement of housing conditions and the quality of public and private space around is of course, also present, as derelict buildings often also contribute to the social stigmatization of neighborhoods. People-orientation is reflected for instance in a needs-based adaptation of apartments and their layouts, the establishment of intercultural gardens, a reorganization of public spaces supporting their multifunctional use and appropriation and the promotion of a social mix of residents (which is, however, difficult to achieve by direct interventions).

A great number of issues relevant to the livability of an area and the opportunities of the local population can potentially be addressed by funding. First and foremost this concerns social and ethnic integration. For this purpose, the program supports measures improving educational facilities and social infrastructure, in particular, taking into account that early childhood education and professional qualification of the socially disadvantaged are crucial for equal opportunities and for integration of minorities. In many cases, close cooperation between urban planning and school departments aims at combining physical and social measures to turn schools into veritable neighborhood centers. For this purpose, school compounds are reorganized, linked to pre-school facilities and opened up into the communities. Ambitious cities establish complex education networks and reform their educational system for better support of children from poor families or ethnic minorities.

To support social and ethnic integration, multicultural centers, youth centers, self-help orientation and self-organized services as well as street festivals are supported. As deprived areas are often hit by economic transformation and unemployment, the program also tries to promote the local (often ethnic) economy. For this purpose, qualification partnerships among local businesses and educational and training facilities are very important. Support for new businesses and ethnic economies is given mainly with the help of and training programs. City marketing activities are focusing on specific profiles of ethnic shopping streets. Beyond that range of measures, ambitious neighborhoods develop strategies in the fields of mobility, environment, culture and sports, public health and security. PR strategies also try to overcome negative stigmatization and promote a better image.

4 Achievements and shortcomings of the socially integrative cities program

If one evaluates the range of activities in the more than 900 designated areas having received public funding in all parts of Germany, it is important to note a great diversity of their application. Due to the fact that local governments can define the goals of their regeneration strategies individually according to the respective challenges they are facing. The mentioned fields of action can be seen as an optional set of sub-strategies that may be funded but can be selected if there is a need for them in the local context. While housing and public space is a major focus of many of the regeneration areas, the major innovation of the socially integrative city may be seen in the field of social and ethnic integration. Program evaluations confirm that funding has often indeed strengthened social networks and initiatives (BBSR & BMUB 2017).

Digging deeper into the institutional setup of people orientation, the program has contributed to empowerment and a substantial increase in the participatory generation of upgrading projects. Major reasons are the empowerment orientation of neighborhood management, the catalytic role of social infrastructure and the opportunities for micro-projects offered by neighborhood funds. These effects have not totally been able to counter the selectivity problem of participation in urban planning, though: people disposing of resources such as social capital and time are generally over-represented not only in general planning participation, but also in the context of the socially integrative cities program (Selle 2013; BBSR & BMUB 2017).

When it comes to assessing the role of institutional innovations, there has nevertheless been clear evidence that they are successfully implemented and appreciated by the local stakeholders. The procedural setup of integrated development concepts is in most cases able to get a variety of municipal agencies and citizens involved: For instance, in 91% of the cases, the department for social affairs participated in the preparation of the integrated development concept directed by the planning department, while the respective figure was 79% for the parks commission and the school department, 78% for the children and youth welfare service and 72% for the traffic authority (Difu 2015; BBSR & BMUB 2017:48). Community-based organizations participated in 88% of all cases, schools in 82%, and housing companies in 78% (Difu 2015; BBSR & BMUB 2017:49). The crucial part for people orientation, however, is the neighborhood management offices that were established in 85% of all cases. While they have a great number of different tasks in managing the regeneration, the key of their work is empowerment: Around 95% of all neighborhood managements are actively empowering and participating in the local population, publicly inform about the regeneration efforts and act as a one-stop office for the residents and coordinate neighborhood action (Difu 2015; BBSR & BMUB 2017:55).

Summing up, the new program tackles urban problems with a stronger social focus in five different yet inter-related ways: First, it improves local living conditions by upgrading the standard and state of the physical environment in a rather traditional way. Second, it improves the livelihoods and opportunities of the local population with the help of better (social) infrastructure and amenities. Although earlier stages of urban regeneration already worked in a similar direction, the socially integrative cities program concentrates much of its efforts in this field and mobilizes extra public funds that are channeled into those activities. Third, it improves the chances for social integration with the help of positive role models and chances for positive social learning. This is of particular importance since socially disadvantaged areas often produce a great number of school drop-outs and adolescents that lack professional orientation, miss career options and might ultimately be lost for better social integration. Fourth, the image of neighborhoods can often be improved to avoid or reduce future negative stigmatization and discrimination. Fifth, capacity building by empowering self-help structures and raising political attention for the socially disadvantaged strengthens the role of disadvantaged areas, possibly allowing for better consideration of their needs in political decisions at a municipal level.

Nevertheless, while urban regeneration in this program often efficiently restores confidence in a brighter future of funded areas, the impact of projects and measures remains somewhat limited due to their insecure financial and institutional status: Urban regeneration funds aim at upgrading and stabilizing the situation in a designated area for a decade or even 15 years. However, after that period a regeneration area is supposed to flourish without additional funding. In the context of socially integrative city areas this expectation is difficult to meet as social networks often require long-term stabilization and support. The physical structures (neighborhood centers, sports facilities and the like) created with that intention do often have positive effects on social and ethnic integration. When one looks at the economic impact, one has to admit that the stabilization needs a long breath and that the success stories are rare in this context. It is mainly the positive image of ethnic shopping streets and the qualification of adolescents that should be noticed here. While selective outmigration of the neighborhoods, once feared as a major driver for their instability and stigmatization, has not really been overcome in the respective areas of economically weaker towns and cities, a general trend towards "re-urbanization" has recently seized many cities, showing that inner-city neighborhoods are again very attractive for yuppies and even ordinary families. In this context, one can note that the socially integrative cities program has contributed to the renewed appeal of many neighborhoods. On the other hand, less affluent cities are still facing severe socio-spatial polarization and local islands of poverty. Despite the stabilizing effect of programs like the socially integrative cities, universal challenges of societies in the transformation towards the post-industrial age can hardly be overcome with the limited funding of area-based regeneration.

5 Conclusions: co-production in urban regeneration

The historic overview has demonstrated a general trend away from top-down planning towards stronger integration of resident populations into the preparation and implementation of regeneration strategies. Early experiences have shown that demolition and renewal often lead either to unattractive and featureless modernist settlements lacking vibrancy or to replacing traditional neighborhoods with elitist luxury condominiums that disrespect the social ties and the limited financial means of the socially disadvantaged. Starting in the late 1960s, more integrative and participatory approaches have shown that more careful regeneration is possible that respects the needs of the local population and strives for livable neighborhoods. The micro-regeneration approaches that gradually adapt historic quarters to contemporary needs build not only on local knowledge but also on a thorough investigation of inner-city living. They have brought about institutional innovations that changed self-conception and increased the importance of participatory planning. After starting with analyses of the potential negative impacts of urban regeneration, resident councils were introduced that had to be heard in the preparation of regeneration strategies. As the focus of intervention switched from physical upgrading to a more integrative approach including socio-economic deprivation and the chances of the socially disadvantaged, neighborhood managements actively promoted empowerment and the generation of micro-projects governed and implemented by the local population.

This general trend is embedded in a major turn in urban planning in Germany and other countries since the end of the 20th century that has been described by theorists of collaborative planning such as Patsy Healey (1997) or, in Germany, Klaus Selle (1994, 2005). In the last two decades, it has produced a number of additional innovative planning approaches that can be subsumed under the heading of "co-production". This notion makes clear that urban space is no longer seen as merely physical but always socially constructed; it is only constituted by human interaction in Euclidean space. This understanding legitimizes recent local initiatives that sail under the flag of the "right to the city" (Lefebvre 1968), claiming that the planning and the use of space in the city should not be limited to lower degrees of participation (Arnstein 1969), but that people have the right to "co-produce" space. Those initiatives prevail in inner cities, since it is there where space is limited and its use is often restricted by ownership rights and private profit interests. Although they are often limited to small areas in the city or only one single plot, they stand for a different understanding of urban regeneration where empowered people develop alternative urban forms of living. Sometimes they have to be seen as the only means to inject new life into derelict areas in stagnating areas cities in post-industrial transformation.

The initiatives come in a variety of forms. One major approach is bottom-up activities by small self-organized community-based groups to save vacant yet historically important buildings like the so-called "Schwabehaus" in Dessau (www.schwabehaus.de) or the "Straze"-project in Greifswald (www.straze.de), both developing non-profit cultural centers open to the local public with a lot of individual dedication. Another one is the transformation and temporary use of vacant open space in shrinking cities for parks, urban gardening or many other activities that create urban life beyond profit orientation (Oswalt, Overmeyer, & Misselwitz 2013, BBR 2004). Recently, this approach has flourished and challenges conventional planning when profit interests are back on private plots (www.prinzessinnengarten.net). Urban interventions show how new life is possible in inner cities and sometimes even kick off new alliances between planners, architects and local residents (Altrock & Huning 2015).

It is important to note that the framework of urban development grants that was established 50 years ago and has since characterized the upgrading and stabilization of declining neighborhoods in maturing cities where no easy profits by private developers are at hand often offers parts of the solution even in the context of co-production. New and experimental forms of urban life often evolve with its support, be it in the context of empowerment as in the socially integrative city or in the context of bottom-up co-production. Thus, urban regeneration and its system of public funding have continued to play a crucial role in this context of inner-city urban development apart from the domain of profit-oriented CBD development.

第二部分

城市历史与保护

从百年金街王府井看城市设计的变迁与未来

The transformation and future of urban design: Taking an example of Wangfujing, a century-old golden street

郭　婧　石晓冬
Guo Jing　Shi Xiaodong

摘　要：北京王府井几十年的探索实践反映了城市设计在首都发展中发挥的作用和历程的变迁。进入新时代，王府井不仅已登上世界商街舞台，同时也是首都的窗口，更是饱含历史底蕴和人民情感的城市发展里程碑。其城市设计建立在尊重历史、深谙文化的基础上，从人的视角出发，反复推敲和尝试，勇于创新和突破，以开放与合作的态度塑造步行者的天堂。在这个过程中，城市设计从宏观到微观全面建立了跟社会现代化治理之间的关系，通过一套涵盖全要素的语言工具，广泛响应了经济发展、社会繁荣、宜居环境和文化活力等多层次的社会需求。

关键词：王府井；城市设计；变迁；以人为本

Abstract: Decades of exploration and practice in Beijing Wangfujing District reflect the changing role and process of urban design in the development of the capital. Entering a new era, Wangfujing is not only one of the world's best commercial streets, but also a window of the capital. It is also a milestone in urban development that is full of historical heritage and people's emotions. Its urban design is based on respect for history and profound culture. From the perspective of human beings, we have repeatedly considered and tried to make breakthroughs, and have shaped the pedestrian paradise with an open and cooperative attitude. In this process, urban design has established a comprehensive relationship with social modernization governance from macro to micro level. Through a set of language tools covering all elements, it has broadly responded to multi-level social needs such as economic development, social prosperity, livable environment and cultural vitality.

Key words: Wangfujing; Urban Design; Change Process; People-Oriented

1　王府井地区的基本情况

王府井大街始建于元代至元四年(1267)，始称哈达门丁字街，定名于清光绪三十一年(1905)，距今已有750余年的历史。明代时期，在此修起了10座王府，王府井也就初具规模，称“十王府街”(图1),至清光绪二十九年(1903),以东安市场(清朝北京最早的综合市场)开业为标志,开启了王府井大街百余年的商业街史。

清末民初的王府井,是外国人了解北京和中国的窗口,是现代教育的生发地,也是中外文化交织碰撞之地。贝满女中、陶氏私立两等小学堂、联合女子圣道学院、辅仁大学、协和医学院,不远处的京师大学堂(北京大学前身)、中法大学等,无不在中国近代教育史上占据重要的篇章;教堂、学校、医院、书馆、戏院、饭店、旅馆,无不为这片具有数百年历史的商业区增添着文化内涵。从商人到政客,从学者到医生,从军官到教师,出入王府井的历史人物们留给中国的人文积淀,在清末民国的动荡中仍熠熠生辉(图2)。

图1　十王府街旧景
(来源:北京市规划院图片库)

图2　清末女子逛王府井大街
(来源:北京市规划院图片库)

中华人民共和国成立后,北京选定王府井大街作为首都“窗口”商业街,北京市百货大楼(图3)、新中国妇女儿童用品商店、北京工艺美术服务部和王府井新华书店相继在王府井大街开业,东安市场进行翻修,中国照相馆、雷蒙西服店、浦五房肉食店、四联理发店(图4)、普兰德洗染店等一批名店由上海迁至王府井大街落户,王府井大街从而成为独步京华、享誉中外的商业名街。

图3　1959年的王府井百货大楼
(来源:北京市规划院图片库)

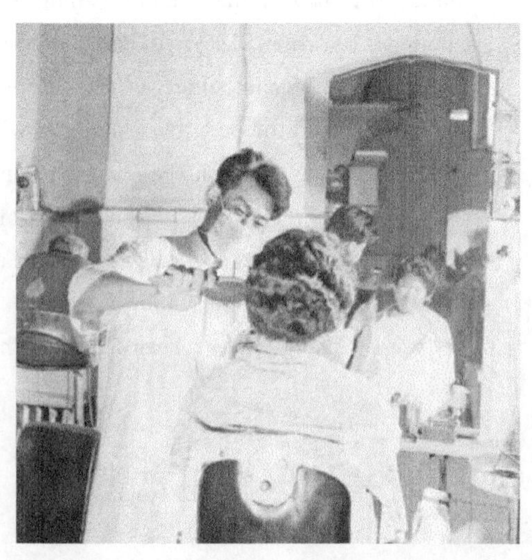

图4　四联理发店自上海迁入王府井
(来源:北京市规划院图片库)

改革开放之后，王府井迎来了更为全面的发展，1999 年 9 月 11 日，王府井作为中国首批商业步行街开街，这个 700 多年的历史老街与法国香榭丽舍大街结成了友好街，作为服务北京全市乃至全国、面向全世界的国际化现代化商业中心掀开了新的一页。

2019 年 12 月 20 日，王府井步行街再次提级，北延段(金鱼胡同至灯市口段)正式亮相，打造成更加国际化、人性化和具有活力的步行街区。

在 700 余年的发展历程中，王府井地区凭借优越的区位和资源，始终处在领先时代的地位，这里有清朝时期北京最早的一座综合市场——东安市场，民国时期京城最优秀的中学——男育英、女贝满，新中国第一店——北京百货大楼，至今国内最顶尖的医学院——协和医学院，等等。同时，王府井还时刻秉持着它特有的文商共享、包容创新的精神，从故宫东华门外商家自行聚集，到积极接受外来的科学教育普及，外来的传教士在此兴建学校、投资医院，这种对不同人群、不同业态、不同文化的包容接纳的态度，与既重商业发展又重文化引入的发展思路相得益彰，促进了区域的繁荣。这种“文商共享、包容创新”的精神，也为其在近现代的发展中积蓄了深厚的力量。

2　王府井发展历程变迁与城市设计的作用

1992 年，北京市政府为引进外资、扩大内需，决定改建王府井，使其成为世界一流的商业中心。随后，东安市场率先进行改建，东方广场、工艺美术大厦、百货大楼北楼、丹耀大厦相继开工，大范围改造很快暴露出一些问题，包括整体市政基础设施改造滞后于建筑改造、改造规划规模过大影响了开发效益、建筑改造缺少城市设计的指导而无法延续传统风貌和塑造整体形象、长期处于施工状态导致市场衰退以及开发管理体制不顺。鉴于此，北京市政府为理顺开发机制，将王府井开发办权力下放，明确不再扩大开发规模，暂停新的建设项目，转变思路，将王府井商业街按照公交步行街进行整治，长远目标为建成完全步行街。由此，开启了王府井商业街的三次城市设计改造提升工程(图 5—图 8)。

图 5　王府井早期改建实景 1
(来源：北京市规划院图片库)

图 6　王府井早期改建实景 2
(来源：北京市规划院图片库)

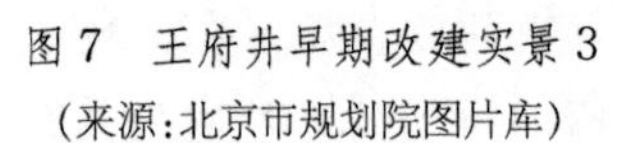

图7　王府井早期改建实景3
（来源：北京市规划院图片库）

图8　王府井早期改建实景4
（来源：北京市规划院图片库）

2.1　第一阶段城市设计：转变大拆大建思路，引领风貌特色塑造

1998年，王府井地区开展了《王府井商业街整体城市设计》，工作重点整体转移到商业街的环境景观整治上。第一，彻底改变市政基础设施的落后状态，一次性完成大街市政基础设施的建设，七种管线全部建设到位，电力电信线全部入地。第二，按照公交步行街的原则管理交通，坚持步行优先，调整断面和路板，路板宽度12米，行道树、路灯等都沿原道路规划16米范围位置布置，这样就在车行道与人行道之间留出了4米宽的一个中间区域，垃圾桶、绿化、座椅等街道家具均布置在这一区域内；人行道与车行道之间不设置路缘石，只是以不同的铺装材质区分，加强商业街的一体性。第三，按照国际一流步行商业街的定位和建设标准，充分体现历史文化内涵和以人为本原则，布置各项环境设施、刻画重要景观节点、整治沿街建筑风貌，北京饭店、百货大楼、东来顺饭庄、穆斯林大厦等具有历史标志性的建筑立面被保护下来，如今我们熟知的百货大楼前广场等就是在这次工作中实现的。第四，塑造独具特色的广告橱窗、夜景照明和环境艺术小品，清理沿街杂乱的广告，追求典雅、大方的效果，突出商业街的主题和特色；以橱窗照明为主导，减少霓虹灯的设置，整体提升商业街的夜景效果；确定王府井商业街的主题形象标志，通过南口牌匾、井盖浮雕设计、写实雕塑等，强化商业街的主题形象。

这项工作是在王府井商业街经过较大规模开发的状况下转变思路而进行的微观城市设计工作，该项工作坚持了统一、人本、文化、简洁四大原则，体现了综合性和系统性。设计方案把国外步行商业街的理念与北京的实际充分结合，通过改善基础设施、把握空间尺度、强化场所精神、营造独特环境，对保留下来的传统商店开展了适度得体的整治，以现代而典雅的环境特征去适应大尺度的空间，不仅使王府井原有的肌理得以延续，更与首都国际化大都市的形象和地位相称；同时，最大限度强化原有建筑特色及空间形态，以南入口处的牌匾、新东安市场附近的写实雕像及施工中发现的“井”，精准展示了地区历史文化特征，既高雅又不乏市井民俗，营造出王府井商业街独特的环境氛围。

1999 年 9 月,王府井商业街正式开街,商业街全长 1 140 米,其中步行街部分长 810 米。开街之后,王府井客流量增加,商业效益增长,老北京的历史文化认同以及区域环境质量得到了提高(图 9—图 13)。

图 9　王府井开街仪式
(来源:北京市规划院图片库)

图 10　开街儿童表演
(来源:北京市规划院图片库)

图 11　百货大楼前广场
(来源:北京市规划院图片库)

图 12　新东安商场门口的雕塑
(来源:北京市规划院图片库)

图 13　王府井开街牌匾
(来源:北京市规划院图片库)

2.2　第二阶段城市设计:传承与展示文化,推动空间品质提升

王府井步行商业街开街后获得全国人民的喜爱,其商业氛围不断提升,接待客流数量不断攀升,进而暴露出一些有待加强之处,例如:商业结构比较单一,缺乏餐饮、娱乐设施,周边交通不够顺畅,缺少绿化休闲广场等。据此,王府井地区又先后开展了两次城市设计提升工作,作为《王府井商业街整体城市设计》工作的延续和完善。

1999 年底,为了实现集购物、饮食、旅游、休闲、娱乐等为一体的多元错位经营格局,努力营造王府井地区的文化氛围,形成该地区文化古都与现代城市特色兼具的优美环境,新一期的城市设计提升工作得以开展。本次工作着重塑造金鱼胡同口、王府井天主教堂广场(图 14)、利生体育馆广场(图 15)、东安门夜市小吃街等重要节点,形成了王府井商业街北延、西进、东扩的“金十字”构架。沿着“金十字”构架,持续改善市政基础设施条件,进一步解决交通与停车问题,为商业和文化氛围营造打好基础;同时,不断丰富商业服务内容,改善历史环境,形成各具特色的空间氛围;并对夜景照明、商业广告、店面橱窗等都要求精心设计,逐项落实,使城市设计的内容渗透到每一方面,切实提高了地区文化品位。

图 14　天主教堂广场
（来源：北京市规划院图片库）

图 15　利生体育馆广场
（来源：北京市规划院图片库）

2000 年底，为了解决步行街建立以来不断增长的交通压力和绿化缺乏问题，又一次城市设计提升工作接续开展。本次工作着眼于区域交通组织需求，打通了王府井西街、校尉胡同 2 条道路，建立交通引导系统，建设停车楼，缓解步行街建立以来不断增长的交通压力；同时，以中国美术馆、金帆音乐厅、人民艺术剧院、商务印书馆、中华书局等为依托，在商业区中融入文化功能，促进王府井商业街的升级；接着，针对地区缺乏绿化的问题，历经 190 天，在王府井商业街西侧，着力建设了皇城根遗址公园。其中，皇城根遗址公园的恢复极为复杂，除市文物局负责挖掘整理皇城遗址工作外，同步进行深化和实施的还有雕塑方案、夜景照明方案、交通规划方案、大街两侧门店及楼房装修美化方案，以及大街两侧违法建筑的拆除、单位及居民搬迁、植树绿化等。2001 年 9 月 11 日，皇城根遗址公园正式开园，如今已成为北京市民、外地游人及外国友人争相参观游览的一处重要场所。

2.3　第三阶段城市设计：依托城市大事件，加强环境整治

奥运会前夕，为满足奥运会比赛要求和城市发展需要，提高城市品质，改善人居环境，北京首次在全市范围内开展了“北京 2468 环境整治”公共空间整治规划，针对北京的两条轴线、四条环路、六大重点区域、八条门户线路，开展了涉及建筑界面、道路交通、绿化植被、市政设施、城市照明等 10 项要素的专项整治行动。

王府井商业区作为六大重点整治区域之一，以问题为导向，针对各类环境要素开展了一系列整治工作，例如，针对严重影响景观风貌的地面市政箱体，开展了小型化、隐形化、美化与绿化设计，净化了公共空间；针对商业街沿街建筑和重要节点，开展了面向大型节庆的夜景照明设计，提升了夜间商业活力；在公共环境艺术方面，以奥运为主题，在多处人流密集节点设置主题雕塑展览，烘托了奥运氛围(图 16、图 17)。

本次城市设计工作，是以大事件为契机，以问题为导向，对王府井商业区开展的环境整治，周期短、可实施性强、效果明显，在一定阶段内取得了良好的治理成效。

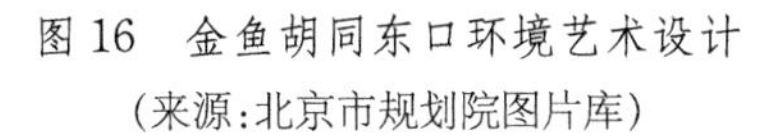

图 16　金鱼胡同东口环境艺术设计
（来源:北京市规划院图片库）

图 17　市政箱体改造后实景图
（来源:北京市规划院图片库）

3　新时代更新治理中的城市设计:强化以人为本,带动街区繁荣

近些年,随着电商冲击和居民消费结构转型,传统商业尤其是百货业更加面临前所未有的压力,北京商业地产进入存量提升时代。王府井也不例外,自其作为步行商业街开街以来,在不断自我更新和生长的过程中,也不免暴露出了一些新的问题。首先,商业街区的道路资源发展受老城整体保护影响,内部交通疏散能力不足,同时周边路网高峰期承载能力已接近饱和,导致街区交通秩序较为混乱,出现了机动车违规停放、非机动车随意停放,挤占步行及骑行空间等问题,步行街周边地区的步行体验较差。其次,王府井商业主街的各大商场各自为战、缺乏衔接发展,且主街与周边街巷尚未实现协同发展,背街小巷的业态杂乱、环境品质低下,街区内部缺乏有机整合。再次,王府井商业街区与其周边的东华门、皇城根遗址公园、美术馆、隆福寺、金宝街等历史遗迹、文化设施缺乏联动发展和整体协调,未能充分发挥文化对街区发展的作用。最后,街区发展的不平衡体现在环境品质上,出现了建筑风貌缺乏引导,垃圾箱、座椅、导引标识等城市家具的布局和设计品质不佳等现象,甚至在局部地区存在私搭乱建和拆违之后留下的城市伤疤。

2018—2019 年,为了更好地引导王府井商业街区的整体提升,激发新的商业活力,北京市政府决定以 8 号线地铁建设为契机,将步行街向北延长,将本区域打造成独具人文魅力的国际顶级步行商业街区。新一轮的综合提升工作以步行街为轴,确定了 1.65 平方千米的街区规划范围。在该街区内,从人员构成上看,有 2 466 家商户、日均约 8.3 万客群以及 3.3 万常住居民(2018 年数据);从空间要素组成上看,有 674 宗权属地块、37 条街巷、18 个大型商业设施以及 420 栋大型公建(2018 年数据)。如何满足街区每类人的需求、布局好每一项空间要素,成为本次工作的重要课题。为此,本次工作从创造以人为本的交通环境、传承饱含情感的人文底蕴、塑造宜游宜赏的宜人街区、刻画多元得体的建筑形象四个角度切入(图 18)。

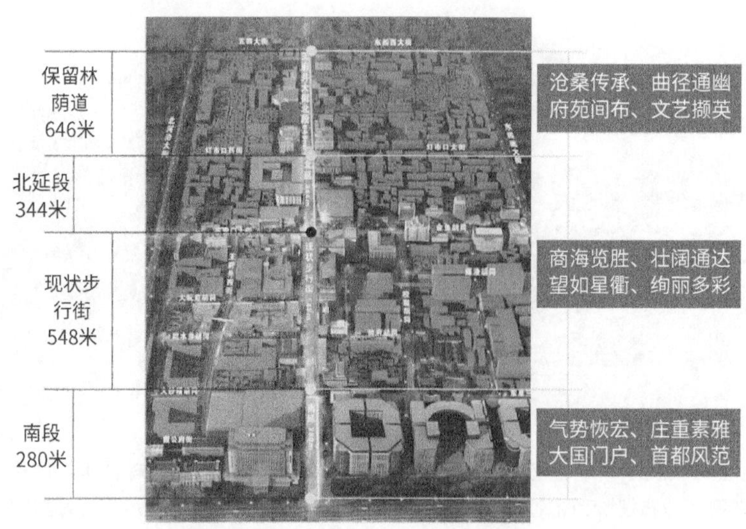

图18　王府井商业步行街向北延长及分段定位示意图

（来源：《王府井商业区综合规划研究》课题组）

3.1　创造以人为本的交通环境

王府井街区位于首都核心区域，内有协和医院、西侧毗邻故宫和天安门，游人、商户和居民的交通组织交织，需求极为复杂，周边干道的交通压力极大。随着步行街向北延长，原有公交线路改线，相应街区内、外多条道路的交通组织压力将会持续增大，居民的出行将受到明显影响。区域交通再组织是街区发展的前提。

首先，本次工作全面梳理了街区机动车交通组织、区域联动调整公交线路，来解决步行街延长带来的问题。通过调整街区整体通行规则，尤其是将金鱼胡同以北地区调整为单行为主的交通组织模式，充分解决了社会停车场、商场货运出入口和居民居住区的可达性，并局部打通了几处断头路，畅通地区交通循环；同时，在更大区域内，联合交通部门制定公交线路导改方案，确保步行街延长后，居民的地面公交出行不受影响，且导改后的承接路段仍在交通承载能力范围内。

其次，以步行街延长为契机，通过静态交通治理创建了"地面无车街区"，全面改善街巷慢行环境。步行街的延长、道路单向交通的组织模式，为街区全面提升慢行环境提供了良好的空间基础，但街区居民的机动车停车供需矛盾仍旧较为突出。本次工作充分利用街区内充裕的社会停车资源（大型商业综合体的地下停车资源），在街道办事处和王府井管委会的共同协调下，9家大型商业设施面向周边居民共享了630个车位，通过"市场灵活定价＋政府财政补贴"的价格机制，周边居民的停车需求全部得以解决，彻底解放了地面空间（图19）。此后，进一步

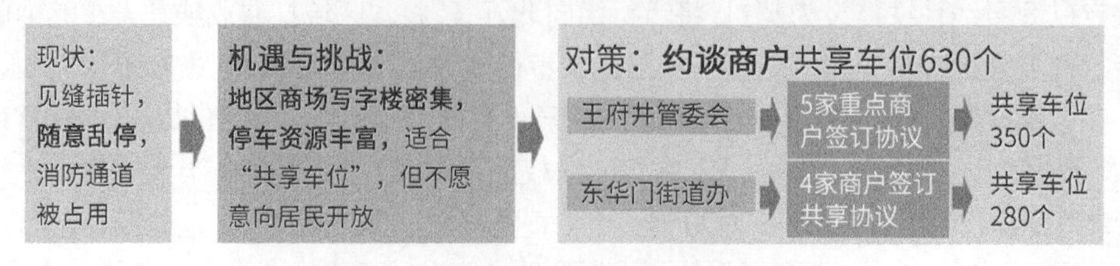

图19　王府井街区共享停车组织模式分析

（来源：《王府井商业区综合规划研究》课题组）

针对韶九胡同、锡拉胡同、北官场胡同、甘雨胡同、柏树胡同、西堂子胡同、煤渣胡同共 7 条胡同进行了环境治理，包括清理违规停车、扩大人行道、增加花草林荫、提供生活服务界面等，全面提升了街区的慢行体验和环境品质(图 20—图 23)。

图 20　煤渣胡同
(来源:东华门街道办事处摄)

图 21　柏树胡同
(来源:东华门街道办事处摄)

图 22　西堂子胡同之一
(来源:东华门街道办事处摄)

图 23　西堂子胡同之二
(来源:东华门街道办事处摄)

3.2　传承饱含情感的人文底蕴

在存量转型发展阶段，文化成为街区商业更新的重要切入点和活力要素。王府井商业街区的文化底蕴深厚、文化类型多元化、文化资源分布较为广泛，因此，以更开阔的视角、更长远的考虑和更严格的措施不断挖掘王府井历史文化内涵，提炼文化价值，传承文化脉络，扩大保护对象，构建其历史文化保护体系非常有必要。

本次工作确定了六大部分保护内容：(1)保护黄图岗传统平房区、报房胡同传统平房区和西堂子胡同传统平房区共 3 片成片传统平房区，以保护区域历史原貌为原则开展更新改造。(2)保护 24 条传统胡同，保护这些街巷的历史尺度和传统风貌，展示其承载的丰富的历史故事和人物。(3)保护 38 处历史遗存，包括 30 处不可移动文物和 8 处历史建筑，在确保安全的前提下，以服务社会公众和彰显文物历史文化价值为导向，鼓励各文保单位和历史建筑用于展览展示、参观游览、文化交流、文化体验服务、非遗传承等使用功能。(4)塑造东安门大街看东华门、帅府胡同看协和医院、嘉德看中国美术馆共 3 个街道对景，在城市设计中予以重点考虑，确保视线通道的畅通。(5)保护好那家花园、翠园、澹园、余园共 4 处历史名园(私家园林)，保护好古树名木 87 棵，其中一级古树 6 棵、二级古树 81 棵。(6)保护好京剧脸谱绘制等 3 项区级非物质文化遗产。

与此同时，为了更好地整合利用和展示文化资源，在街区内设计了多条文化探访线路，在其沿线挖掘历史场景和故事，构建层次丰富的文化景观，并结合首都剧院、中国儿童剧院、老舍纪念馆等重要文化设施以及商务印书馆、张秉贵纪念馆等特色文化资源点，营造界面开放、氛围活跃的文化场所(图 24、图 25)。

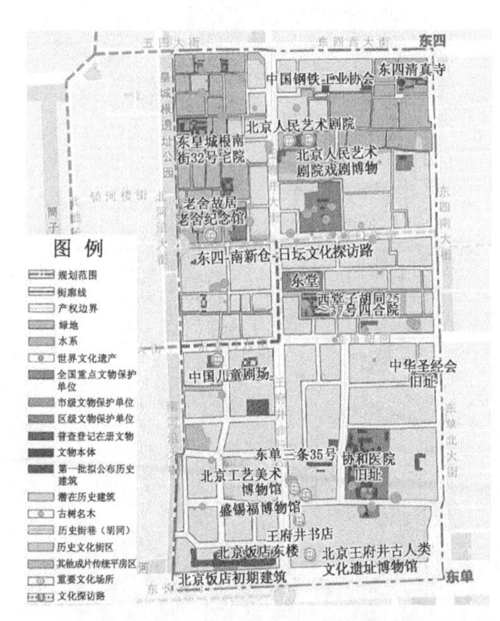

图 24　王府井历史文化保护体系构建

(来源:《王府井商业区综合规划研究》课题组绘制)

贝熙业旧居：

位于大甜水井胡同22-26号院，有一座民国时期的青砖二层、两面坡顶小楼，现存24号门牌。资料记载，法国贝熙业(Bussiere)医生曾寓此处。贝熙业医术精湛、热心中国抗战事业，为抗战胜利做出了贡献。松沪抗战总指挥、新中国首任纺织部部长蒋光鼐也曾居于此。此处有水井一口，井口有题字，据传此井即为“大甜水井”之井。此处楼房建筑保存较好，具有民国时期建筑特征。

图 25　贝熙业旧居的文化故事挖掘和文化场所营造

(来源:北京市规划院图片库,《王府井商业区综合规划研究》课题组绘制)

3.3　塑造宜游宜赏的宜人街区

鉴于王府井商业街区所暴露出来的缺少绿化空间、休憩空间和文化活动空间的关键问题，本次工作从每一栋建筑、每一处地块入手，挖潜存量空间。

第一，塑造绿意环绕、绿荫遍地的公共休闲空间。在传统平房区，结合胡同门前绿化、院落绿化和大树浓荫，以点带面提升绿视率；在背街小巷治理中，结合拆违和拆迁腾退空间，"见缝插绿"，增加邻里交往空间和街区开放休闲空间；针对剧院、美术馆、图书馆、大型酒店和商务办公楼等大型公共建筑，激活建筑前空间，形成散布于街区的丰富而充满意趣的绿色公共空间；在多产权的商业集中区，打破产权界限的局限性，加强协同合作，塑造庭院空间和休闲外摆，提升商业活力(图 26、图 27)。

图 26　王府井校尉胡同西侧拆违前
(来源:《王府井商业区综合规划研究》课题组)

图 27　王府井校尉胡同拆违后改造为口袋公园
(来源:《王府井商业区综合规划研究》课题组)

第二，营造随处可憩、悠然闲适的外摆座椅空间。结合街区内的绿地广场设置外摆座椅，营造动静相宜的舒适氛围；结合酒店、餐厅设置外摆座椅，提供舒适的室外消费空间；结合重要建筑的屋顶平台和室外挑台设置外摆座椅，提供舒适悠闲、视野独特的观景平台。

第三，创造丰富多元、主题鲜明的文化活动空间。鼓励商业主体利用室外场地，开展各项节庆和文化活动。

3.4　刻画多元得体的建筑形象

街区内的建筑历史悠久，其建设年代不一，建设风格极为多元，很多建筑都是其所在年代的典型代表，如北京饭店、中国钢铁工业协会大楼、百货大楼等，更有和平宾馆是杨廷宝大师在京留下的重要建筑作品。这些建筑往往体现了这个时期的建筑风貌特征和时代精神，具有保留和维护的价值。然而，在长期的商业运营过程中，很多建筑的立面被广告、屏幕所遮挡，建筑风貌无处展现，使王府井丧失了这一独特的文化展示界面。

本次工作首先对王府井商业步行街两侧的建筑原貌进行挖掘和考证，确定了具有保留价值的建筑名录，并明确了每栋楼的广告牌匾整治、立面提升和业态升级要求，制订了"一楼一策"工作计划。通过这一举措，不仅最大限度保留了多元的建筑风貌格局，还使整条大街更加富有统一性和整体感(图 28—图 30)。

图 28 百货大楼历史风貌实景
（来源：北京市规划院图片库）

图 29 百货大楼改造前实景
（来源：王府井管委会）

图 30 百货大楼改造后实景
（来源：《王府井商业区综合规划研究》课题组）

3.5 制定街区更新管理规则

王府井街区的更新提升工作涉及历史文化、交通治理、公共空间、绿化景观、基础设施和建筑风貌等，是一个持续而交替进行的过程，建立全周期的更新管理至关重要。为此，本次工作面向管理制定了街区更新设计图则，统筹协调公共空间更新，形成“一楼一策”工作指导方案。本套图则作为管理依据和支撑文件，是对各项工作的归纳、整合和“翻译”（“翻译”为管理语言），是推动规划设计成果实施落地的重要保障，对于管理者、运营者、社区居民以及后续加入的专业设计人员等的共识凝聚至关重要(图 31)。

4 结语

北京王府井几十年的探索实践反映了城市设计在首都发展中发挥的作用和历程的变迁，从塑造特色风貌、提升空间品质、加强环境整治，再到回归以人为本的综合治理。进入新时代，王府井不仅已位列世界商街舞台，同时也是首都的窗口，更是饱含历史底蕴和人民情感的城市发展里程碑。其城市设计建立在尊重历史、深谙文化的基础上，从人的视角出发，反复推敲和尝试，勇于创新和突破，以开放与合作的态度塑造步行者的天堂。

在这个过程中，城市设计从宏观到微观全面建立了跟社会现代化治理之间的关系，通过一套涵盖全要素的语言工具，广泛响应了经济发展、社会繁荣、宜居环境和文化活力等多层次的社会需求。这套语言工具，对于政府起到了全周期的控制、引导、评估和监督作用，对于运营者和专业设计人员则起到了能力提升、摸清路径的作用。这恰恰是新时代城市设计的核心价值所在。

图 31　王府井商业街区更新图则示例

(来源:《王府井商业区综合规划研究》课题组)

参考文献

[1] 王引,石晓冬.浅议商业街整治规划与实施:以北京王府井商业街整治规划为例[J].规划师,1998,3(14):51-54.

[2] 北京市城市规划设计研究院.王府井商业街整治城市设计[Z].北京:北京市城市规划设计研究院,1999.

[3] 北京市城市规划设计研究院.王府井商业街(二期)整治工程规划设计[Z].北京:北京市城市规划设计研究院,2000.

[4] 北京市城市规划设计研究院.王府井地区三期整治工程规划成果汇编[G].北京:北京市城市规划设计研究院,2001.

[5] 董光器.对王府井商业街建设与整治的认识[J].北京规划建设,2001(4):18-20.

[6] 王之鸿,姚德仁.王府井[M].北京:北京出版社,2005.

[7] 北京市城市规划设计研究院.北京市重点大街重点地区环境建设概念规划、综合及实施[Z].北京:北京市城市规划设计研究院,2008.

[8] 北京市城市规划设计研究院.王府井商业区综合规划研究[Z].北京:北京市城市规划设计研究院,2020.

广州近现代工业遗产调查与保护再利用

Investigation, protection and reuse of modern industrial heritage in Guangzhou

吕传廷　周建威

Lyu Chuanting　Zhou Jianwei

摘　要：广州不仅是中国近代重要的工业城市之一，也是全国改革开放的实验地，具有相当长的工业历史，遗留下来了众多的工业遗产，特别是在白鹅潭地区。伴随城市的快速发展，在"退二进三"政策背景下，白鹅潭片区旧工业区和旧工厂逐渐退出历史舞台，大量的旧工业建筑和旧工业地段整体风貌不断遭到破坏，其中大量具有历史、艺术、科学技术、经济和文化价值的工业遗产正快速从城市里消失。如何在新的背景条件下，抓住时机，充分挖掘其内在的潜力和开发的价值，保护近现代工业遗产与风貌，成为广州城市建设的当务之急。本文重点通过对白鹅潭片区各阶段的规划研究分析，探讨了该地区的工业遗产保护和发展模式，以期达到工业遗产保护和地区活力复兴的整体最优目标。

关键词：工业遗产；保护再利用；广州

Abstract: Guangzhou is not only one of the important industrial cities in modern China, but also the experimental site of the country's reform and opening up. It has a long industrial history and rich industrial heritage, especially in the Bai'etan area. With the rapid development of the city, under the background of the policy of "Suppressing the second industry and developing the third industry", the old industrial area and old factories in Bai'etan area gradually withdrew from the historical stage, and a large number of old industrial buildings and the overall style of the old industrial area were continuously destroyed. The industrial heritage of historic, artistic, scientific and technological, economic and cultural values is rapidly disappearing from the city. Under the new background and conditions, how to seize the opportunity to fully tap its inherent potential and development value, and protect the modern industrial heritage and style, has become a top priority for Guangzhou's urban construction. This article focuses on the planning research and analysis of the various stages of the Bai'etan area, and discusses the industrial heritage protection and development model of the area, in order to achieve the overall optimal goal of industrial heritage protection and regional vitality revival.

Key words: Industrial heritage; Protection and reuse; Guangzhou

工业遗产保护运动开始于20世纪60年代的英国。国际工业遗产保护协会(The International Committee for the Conservation of the Industrial Heritage, TICCIH)于2003

年通过的旨在保护工业遗产的《工业遗产之下塔吉尔宪章》对工业遗产的界定是:具有历史价值、技术价值、社会意义、建筑或科研价值的工业文化遗存,包括建筑物和机械、车间、磨坊、工厂、矿山以及相关的加工提炼场地、仓库和店铺,以及生产、传输和使用能源的场所、交通设施,除此之外,还有与工业生产相关的其他社会活动场所,如住宅、宗教朝拜地或者教育机构。[1]中国工业遗产保护论坛在2006年通过的《无锡建议——注重经济高速发展时期的工业遗产保护》(简称《无锡建议》)中,工业遗产被进行了精确的定义。[2]工业遗产包括以下内容:具有历史学、社会学、建筑学和科技、审美价值的工业文化遗产,包括建筑物、工厂车间、磨坊、矿山和相关设备,相关加工冶炼场地、仓库、店铺、能源生产和传输及使用场所、交通设施、工业生产相关的社会活动场所,以及工艺流程、数据记录、企业档案等物质和非物质文化遗产。

1 背景

2006年4月,首届中国工业遗产保护论坛在无锡召开,并通过了《无锡建议》,提出要加强工业遗产保护,同年5月国家文物局下发《关于加强工业遗产保护的通知》,特别指出"工业遗产保护是我国文化遗产保护事业具有重要性和紧迫性的新课题",并在2007年第三次全国文物普查中正式将工业遗产纳入调查范围。[3]

2010年11月,中国科学技术协会委托中国城市规划学会开展"典型城市工业遗产保护与科普开发研究"。该课题通过对12个典型城市的工业遗产进行深入细致的调研,试图建立一个工业遗产价值评估标准,对于工业遗产保护和科普开发利用所面临的政策、体制、机制、技术等方面的问题进行深入探讨,在此基础上提出相应的政策措施。[4]在该背景下广州首次对全市范围内工业遗产进行全面的普查。[5]

2 广州工业遗产调查

2.2 广州工业遗产保护历程

广州工业发展的历史中,遗留下来的工业遗产是城市的一笔重要财富,它们见证了广州城市的发展和变迁。广州市"退二进三"和"三旧改造"政策的陆续出台,为广州工业遗产的保存和再利用提供了历史条件及政策依据,为工业企业提供了一个更为经济的处理方式。由于政策规定原有建筑物不得进行房地产开发,大批工业建筑避免了被拆除的命运。对旧厂房的利用,不得拆除外墙、不得改变建筑外轮廓、不得扩建和加建的约束条件,使工业遗产的物质载体得到了较为完整的保存。对发展第三产业的鼓励和优惠措施,使大批工业地被改建为创意产业园和商业地产。其中,信义会馆将广东省水利水电机械制造厂老厂区改造为由政府牵头、开发商主导的综合性商业地产,成为最早的政策示范工程。广州对工业遗产保护的过程大致可划分为三阶段。[6]第一阶段:2004年之前,大规模旧城改造,推行拆除旧厂房,建设居住区。第二阶段:2004年到2012年,出台《关于推进市区产业"退二进三"工作的意见》,明确对旧厂房建筑的利用，以业主自主保护为主，并对发展第三产业做出鼓励和优惠措施，使大批工业地改建为创意产业园。当时形成了一批如红砖厂、珠江啤酒厂、1850 创意园、太古仓、TIT、信义会馆、广州轻纺交易园、农民工博物馆等创意园。第三阶段：2012 年至今，广州市政府出台一系列政策，探索以政府主导的工业遗产保护方式，如广钢新城、广州金融城等规划。[7]

在 1993 年，广州市将广东造币厂纳入广州市第四批法定文化保护单位，这是广州首次将工业遗产纳为文物保护单位。在 2008 年将柯拜船坞遗址纳入第五批市级文物保护单位，在 2008 年广州市"退二进三"和"三旧改造"政策出台后，同年有协同和机器厂旧址、同盛机器厂旧址、近代洋行仓库和码头旧址等 12 处工业遗产被纳入第七批市级文物保护单位(表 1)。

表 1　广州市工业遗产法定文物保护单位一览表

年　份	批　次	名称	保护级别
1996 年	第四批	广州大元帅府旧址(原广州士敏土厂)、广州沙面建筑群(HK 牛奶公司制冰厂、亚细亚火油公司旧址)	国家级文物保护单位
		广东造币厂旧址	市级文物保护单位
1999 年	第五批	柯拜船坞遗址	市级文物保护单位
2008 年	第七批	协同和机器厂旧址、同盛机器厂旧址、近代洋行仓库和码头旧址(包括亚细亚龙唛仓旧址、亚细亚花地仓旧址、渣甸仓旧址、德士古油库旧址、日清仓旧址、美孚仓旧址、太古仓旧址、大阪仓旧址)、五仙门发电厂旧址、广东饮料厂旧址、	市级文物保护单位

资料来源：根据广州市文物保护单位自制

2.1　广州工业遗产普查

工业遗产是城市工业发展过程中形成的工业遗存，调查为城市工业遗产的筛选、价值认定提供了依据。[8]通过广泛的文献资料查阅和分析，构建广州市潜在工业遗产的清单，包括 1949 年以前形成的传统手工工场遗址、民族工业企业、官办或者官商合办以及外资工业企业和 1949 年建国之后形成的重要工业企业。总体来说，主要从以下几个方面构建广州市潜在工业遗产清单，作为调查的对象：(1)各级文保单位中的潜在工业遗产；(2)历次文物普查中登记的潜在工业遗产；(3)历史文化名城保护体系中的潜在工业遗产；(4)其他线索下的潜在工业遗产。[9]

针对近年来工业企业大量搬迁的情况，潜在工业遗产的保护和使用情况发生了巨大变化，一部分有重要价值的潜在工业遗产被拆除，工业遗产调查的主要任务是明确其现存情况，根据工业遗产调查表从总体情况、建构筑物情况、工业设备情况以及文档图纸情况等四个方面对工业遗产进行记录、整理和分析，普查的内容包括工业企业的基本情况、工厂发展历史、建设情况、现状保护和利用情况以及文献档案等。调研期间，在近现代广州市潜在工业遗产清单中共梳理出工业遗产 83 处，调研发掘白鹅潭地区存在近现代工业遗产有 29 处左右，这里面除了保留着完整的洋行码头文化遗产外，还有广州工业品牌的"老字号"。[10]

2.3　广州工业价值评价

广州市工业遗产的价值要以其所处的历史时期为背景，以其工业特征为基准进行认定。

首先,基于对广州市工业遗产的全面系统的调查,对每一类价值进行细化,通过二级评价指标确定每一项价值指标的衡量标准。根据价值评价的分值体系,立足于深入、有针对性的调查,以城市和工业发展为背景,基于同一项价值指标进行工业遗产之间的纵向比较,明确工业遗产之间的价值梯度,区分其重要程度和稀缺性。工业遗产的综合评价应该立足在工业遗产保护和再利用的基础上,考证工业遗产的多重价值。根据工业遗产的综合评价需要,建立了如下评价体系结构:5 个一级指标、10 个二级指标。据此评价各项指标对工业遗产综合价值的影响程度,划定每个指标所占的分数,作为划分工业遗产等级的依据(图 1)。

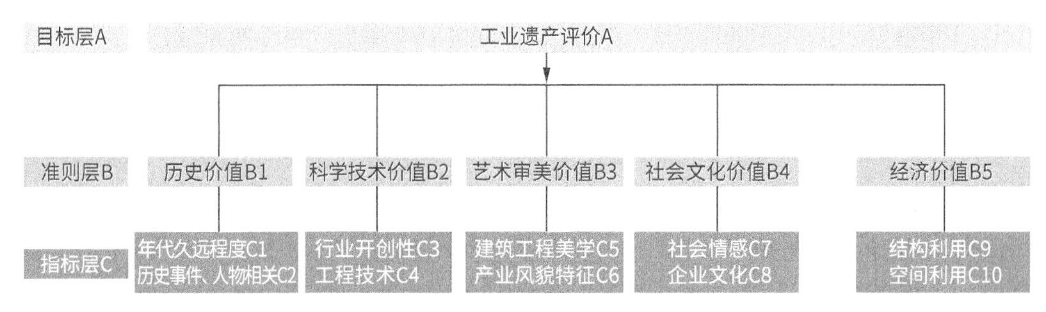

图 1 广州工业遗产价值评价体系

(1) 历史价值指标

历史价值是构成一般文化遗产的重要因素,体现工业遗产作为文化遗产的根本属性。历史价值指标中包含年代久远程度和历史事件、人物相关两方面内容。年代久远程度赋予工业遗产珍贵的历史价值,是记录一个时代经济、社会、工程技术发展水平等方面的实物载体,年代的久远,在某种程度上令工业遗产具有稀缺性。历史事件、人物相关方面令工业遗产具有特殊历史价值。

(2) 科学技术价值指标

该指标用于衡量工业遗产所蕴含的工业技术价值大小。其下分行业开创性和工程技术两个二级指标。行业开创性指某一工业门类、技术、设备应用在同行中具有开创性,从而使这些企业和建筑、设备将具有特殊遗产价值。工程技术指工业遗产的生产基地选址规划、建构筑物的设计、机械设备的安装等应用了当时的先进技术,使工业遗产在工程方面具有科学技术价值。

(3) 艺术审美价值指标

艺术审美价值包括建筑工程美学与产业风貌特征两个二级指标。建筑工程美学指建构筑物的风格、流派、特征、色彩、材料等形式意义表现出来的审美价值;产业风貌特征指工业遗产在形成过程中所留下的独特产业风貌,依靠其地理位置和独特的外形和结构特点,对城市景观和建筑环境产生的艺术作用,具有重要的景观和美学价值。

(4) 社会文化价值指标

工业遗产的社会文化价值指标用来描述社会对工业遗产的集体记忆,包括社会情感与企业文化。工业遗产见证了人类工业化进程中的巨大变化,在社会生活方面起到了引领和改善作用,公众已将社会责任作为衡量企业价值的标准。企业文化是工业遗产价值中所包含的非物质遗产部分,包括企业在经验管理、科技创新等方面的内容,以及流传下来的企业文化、精神、理念等。

(5) 经济价值指标

经济价值主要指工业建构筑物改造和再利用的潜力，包含结构利用和空间利用两方面。这是决定工业建构筑物能否录入遗产名录的最后一项考虑因素。结构利用指工业遗产的建构筑物一般都相当牢固，在对工业遗产的再利用过程中可以节省大量的拆迁和建设成本，避免资源的浪费。空间利用指工业建筑大多具有大跨度、大空间、高层高的特点，其建筑内部空间具有使用上的灵活性，具有极强的艺术表现力和巨大的经济价值。

3 白鹅潭片区工业遗产保护历程

白鹅潭地区位于广州市中心城区西部，与佛山毗邻，是广州市近现代历史文化风貌区(图 2)。白鹅潭片区是广州近代工业文明最主要的遗存地，更是鸦片战争之后中国近代工业发源地与核心遗产之一，滨江一带延绵数里的旧产业区见证了中国近现代工业发展的历史，是城市记忆和文化沉淀深厚的特殊地段，是广州历史的演变和城市空间发展的组成部分，具有极高的历史价值、文化价值、艺术价值和经济价值。白鹅潭片区港口码头作为世界重要港口的历史可追溯至 2 000 多年前，它是古代海上丝绸之路的发源地，也曾是近代中国对外贸易的唯一口岸。在这次广州工业遗产调查中，发现白鹅潭片区存在 29 处潜在近现代工业遗产，是广州工业遗产重要集聚区(表 2)。

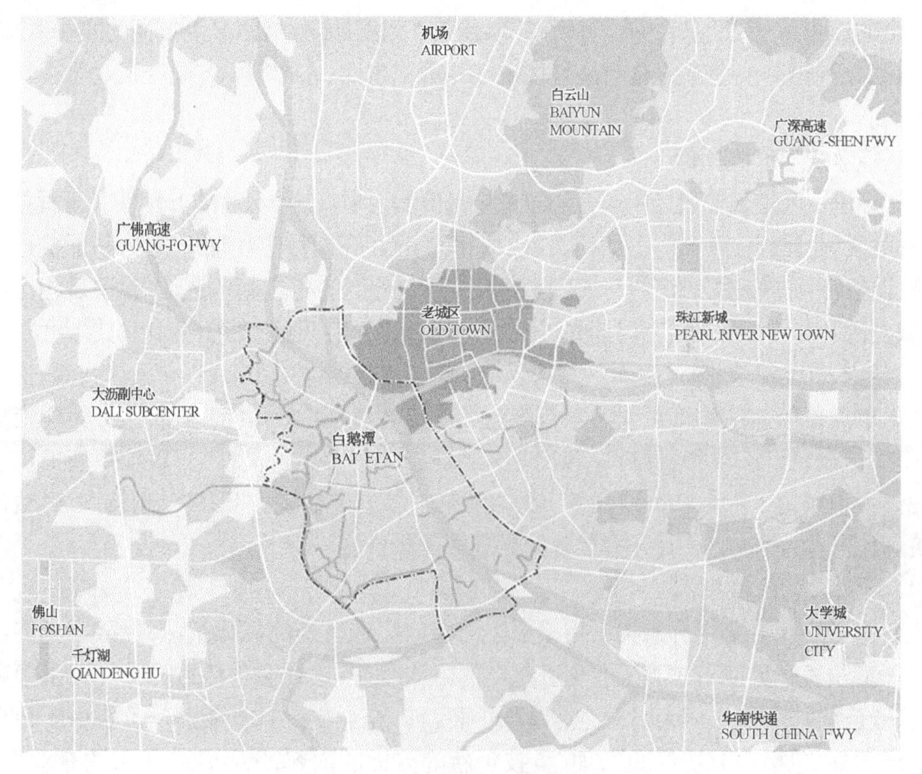

图 2 白鹅潭在广州区位

表 2　白鹅潭片区潜在工业遗产

编号	名称	历史意义
1	广州铝材厂	“前进牌”老字号
2	广州市金源食品厂	广州市唯一以生产南乳、腐乳为主的企业,“雄鸡牌”老字号
3	广州珠江钢琴集团有限公司	世界上最大的钢琴厂商,“珠江”牌钢琴老字号(广州钢琴厂)
4	广州柴油机厂	中国柴油机制造业最悠久的企业
5	广州织金彩瓷工艺厂	“广彩”老字号
6	广州制漆厂	“电视塔”牌老字号
7	新风港滨江仓库	特色工业建筑
8	广铁南站滨江仓库	粤汉铁路遗址、百年车站
9	石围塘滨江仓库	广三铁路遗址、百年车站
10	信义会馆	现状信义国际创意园
11	金珠江化工厂	现状 1850 创意园
12	富民公司储备库	特色工业建筑
13	市粮食局杏村仓库	特色工业建筑
14	五羊自行车厂厂房	“永久”自行车,“老广”的永久追忆
15	广纸滨水区厂房仓库	广州历史悠久的名牌老厂
16	广州钢铁厂	珠三角唯一的钢铁厂
17	广州造船厂	珠三角最大的造船厂
18	省百货凤凰仓及周边	本土老牌企业历史,特色工业建筑
19	昌岗油库	特色工业建筑
20	同盛机器厂旧址	特色工业建筑
21	协同和机器厂旧址	特色工业建筑
22	近代洋行仓库和码头旧址	龙唛仓旧址、花地仓旧址、渣甸仓旧址、德士古仓库旧址、日清仓旧址、美孚仓旧址、太古仓旧址、大阪仓旧址及相关的码头旧址

资料来源:根据相关资料和调研整理

鸦片战争后,从 19 世纪末到 20 世纪的 40 年代,以英美为代表的西方现代工业革命带来的工业技术首次输入中国,在白鹅潭集中建设了大量的外国租界、洋行、码头(仓储),形成了中国近代工业史上第一批集中工业区和华南地区工业制造中心。这些近代工业、交通运输业的历史痕迹至今尚存,记载着昔日对外贸易和工业繁荣与发达的景况,是中国近代对外交流、广州近现代内河港口建设和各类行业经济孕育及发展的重要区域，也是广州早期城市化的代表性地区。该地区在近现代时期对城市形态、经济成长和产业布局有着巨大影响。目前分布在两岸地区的仓库码头旧址有西岸的英商亚细亚花地仓码头仓库、日商日清仓、英商怡和洋行渣甸仓码头仓库、美孚公司仓库、英商亚细亚龙唛仓，东岸的日商大阪株式会社码头仓库、英商太古轮船公司码头仓库(图 3)、美商德士古油库等 9 个洋行码头仓库遗址。

自 1949 年至 20 世纪 90 年代，白鹅潭一带的大部分区域都作为广州市重要的重工业区、货运仓储和水运码头区使用，其历史作用举足轻重。随着城市产业形态的调整和转移，沿江岸线的大部分工厂企业已陆续停产、搬迁。白鹅潭地区保留了以广州柴油机厂为代表的广州工业“老字号”以及以广州钢铁厂(图 4)、广州造船厂、广州造纸厂为代表的现代大型工业文明，很多工业厂房都具有特殊的历史背景与文化价值。

图 3　太古仓码头

图 4　广州钢铁厂内部高炉

历史遗产保护是白鹅潭地区规划提倡可持续发展的一个重要组成元素。由 2004 年开始编制的《珠江后航道两岸近代洋行码头仓库历史遗址保护与利用规划研究》到 2009 年《白鹅潭地区城市设计及控制性详细规划》，再到 2013 年《广钢新城控制性详细规划(修编)》，最初对工业遗产主要是以基础调查为主，在后面规划中逐渐对工业遗产提出详细方案，并提出通过更新策略达到完整保护设想。通过三阶段规划的不断深入，白鹅潭片区工业遗产价值逐步被发现、评估和保护，并逐渐进入保护再利用的过程。

3.2　第一阶段：首次对白鹅潭珠江后航道工业遗产进行摸查

在 2006 年的《珠江后航道两岸近代洋行码头仓库历史遗址保护与利用规划研究》中，首次以文物保护单位为线索对白鹅潭地区近代洋行建筑工业遗产进行调查，并提出再利用的初步设想。在原有规划的基础上进一步挖掘有价值的文化线索，在文化局的协助下，不断丰富历史文化内涵。

(1) 对后航道工业遗产进行全面普查确认、登记建档

工业遗产普查的目的是全面掌握工业遗产的分布、数量、特征、保存现状、环境状况等基本情况，更重要的是通过调查、评估工业遗产的社会、经济、历史价值，将重要工业遗产及时认定公布，实行分类保护，丰富文化遗产的多样性和完整性。随着珠江后航道工业逐渐衰退，许多

有价值的工业遗产在企业停产关闭后面临大量珍贵档案流失的问题。开展工业遗产普查、摸清家底、掌握情况和做好登记认定工作,已成为加强工业遗产保护工作的首要任务。普查对象是全市各个历史时期工业发展阶段遗存的具有典型性、代表性的各类遗产。在规划中,首次对白鹅潭珠江后航道片区工业遗产进行全面普查确认、登记建档,建立工业遗产资料数据库,确定一批亟待抢救保护的项目清单(表3)。

表3　珠江后航道亚细亚花地仓工业遗产普查档案

<table>
<tr><td colspan="2">建筑名称</td><td colspan="2">亚细亚花地仓</td><td>地址</td><td colspan="2">芳村大道东墩头西街</td><td colspan="2">—</td><td>—</td></tr>
<tr><td colspan="2">房屋产权</td><td colspan="2">广州市石油公司</td><td>使用功能</td><td colspan="2">储油</td><td colspan="2">建筑层数</td><td>1</td></tr>
<tr><td colspan="2">建筑年代</td><td colspan="2">1918年</td><td>建筑结构</td><td colspan="2">砖</td><td colspan="2">建筑类型</td><td>仓储</td></tr>
<tr><td colspan="2">建筑质量</td><td colspan="2">良好</td><td>建筑风貌</td><td colspan="2">一类</td><td colspan="2">占地面积</td><td>—</td></tr>
<tr><td colspan="2">建筑保护等级</td><td colspan="2">市级登记文物保护单位</td><td>保护与整治方式</td><td colspan="2">—</td><td colspan="2">—</td><td>—</td></tr>
<tr><td colspan="3">现状照片01</td><td colspan="4">现状照片02</td><td colspan="3">现状照片03</td></tr>
<tr><td colspan="3"></td><td colspan="4"></td><td colspan="3"></td></tr>
<tr><td colspan="3">现状照片04</td><td colspan="4">现状照片05</td><td colspan="3">现状照片06</td></tr>
<tr><td colspan="3"></td><td colspan="4"></td><td colspan="3"></td></tr>
</table>

(2) 对珠江后航道两岸工业遗产进行价值评估

在历史价值方面,工业遗产的保护和保留,对于解释和印证历史事件、传递历史信息,具有特殊的意义。后航道两岸的仓库码头区整体形成年代集中于20世纪初至20世纪40年代,与沙面集中的洋行区具有直接的联系。作为直接的见证者,其对于研究广州乃至全国近代对外开放史、广州近现代内河港口发展史、广州近现代行业经济发展史都具有很高的价值(图5)。

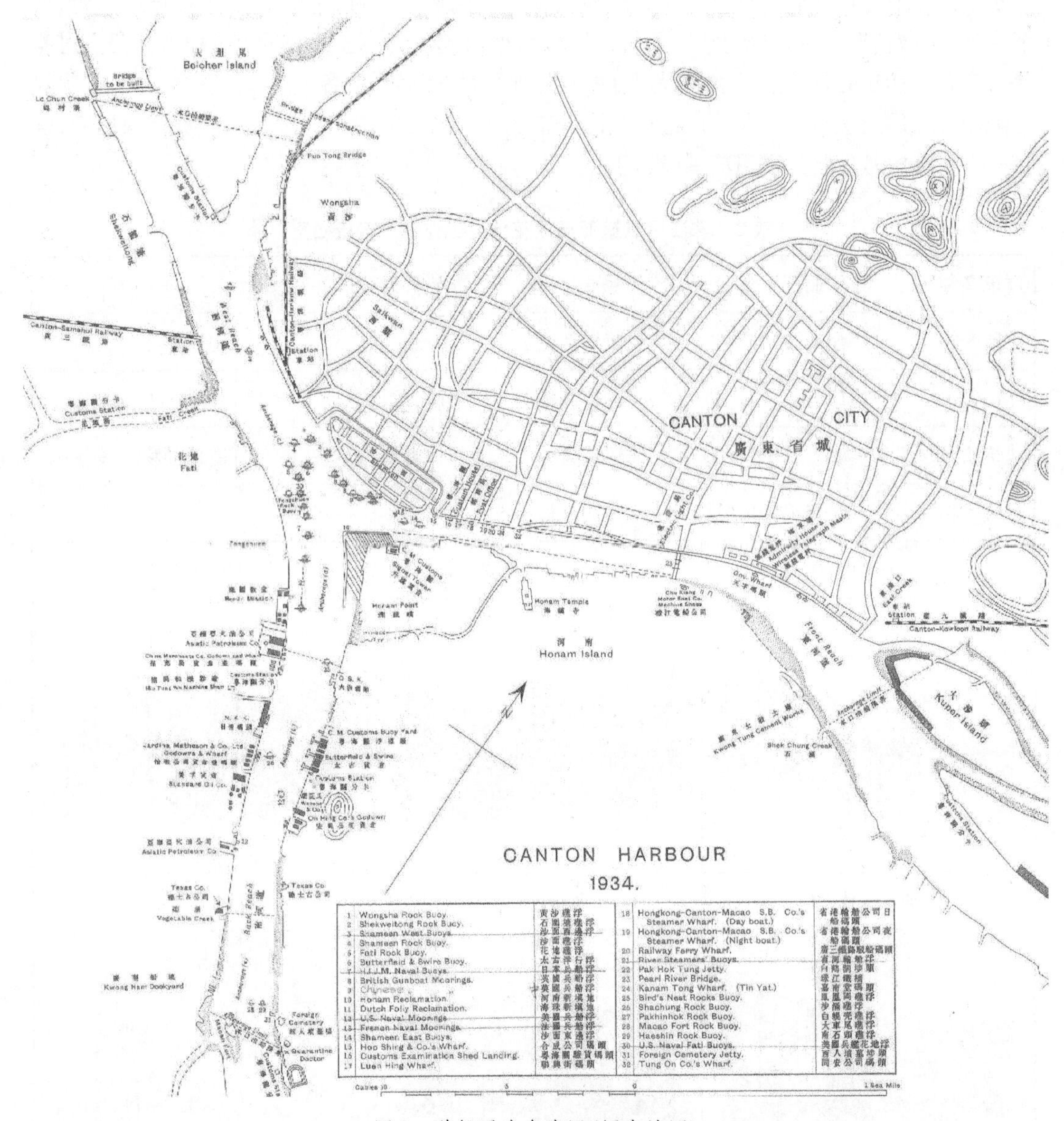

图 5　洋行码头仓库区(历史地图)

在文化价值方面,文化价值具有非物质性,主要是指反映本区几十年发展历史中的重大历史事件,以及存在于几代人记忆中的印象、生活和工作在这里的人们的情感认同。如"港省大罢工""知青乘船赴海南""制造出中国第一台柴油机"等事件,承载了广州人和中国人的自豪。

在艺术价值方面,洋行码头仓库区的建筑多在 20 世纪早期建成,由外国建筑师设计建造,明显具有现代主义和古典主义两种风格交织的痕迹。同一地区、相似的时间内建造的这一批建筑呈现出多重风格的叠加,是研究当时建筑艺术思潮的典型案例(图 6)。

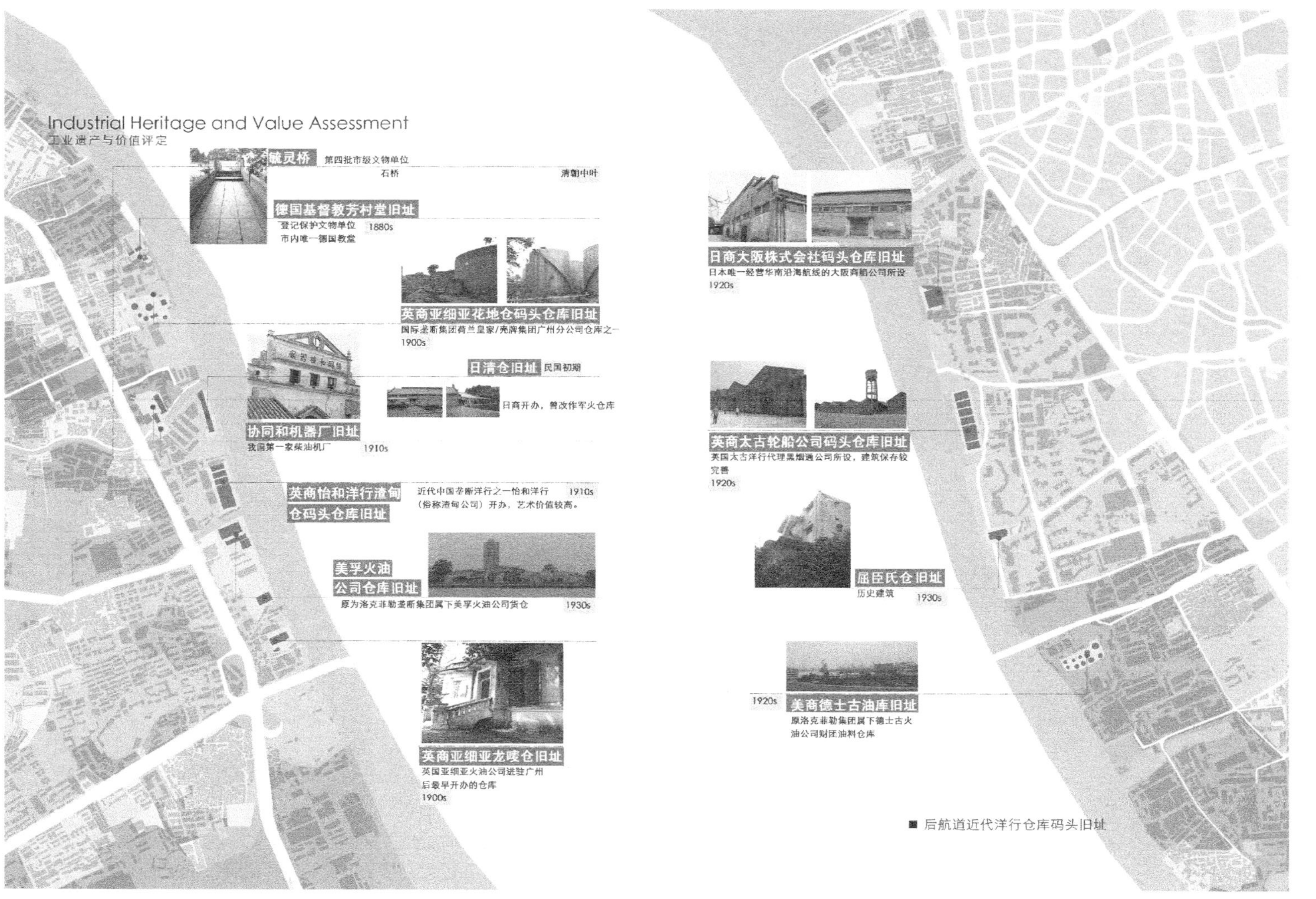

图6 珠江后航道工业遗产与价值评定

(3) 体现分区控制原则，划分工业遗产保护地带

根据区域特征及《广州市历史文化名城保护规划》对历史城区外历史文化街区的相关规定，在区域层面上可将规划区工业遗产划分为工业遗产保护区核心保护区、工业遗产保护区建设控制地带(图 7、图 8)，分别采取不同的保护要求。在详细调查和评估后，将规划区内的优秀建筑分为两级进行保护：一级为文物保护单位，二级为优秀历史建筑。划定文物保护单位的保护范围和建设控制地带，按照文物保护法的要求进行保护。对于工业类优秀历史建筑，可采取相对灵活的保护方式，在保证建筑的立面、结构体系和高度不变的前提下，部分可做一定的创意性设计。

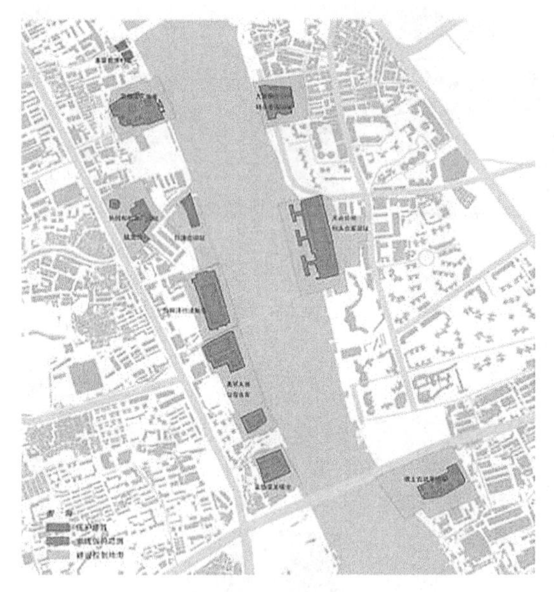

图 7　规划区工业遗产保护单位

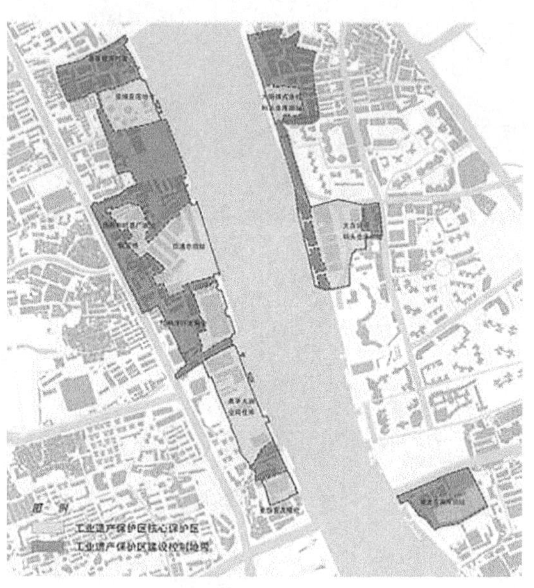

图 8　规划区工业遗产保护区建设控制地带

3.3　第二阶段：保护与更新策略

2008 年出台的《白鹅潭地区城市设计及控制性规划》提出了白鹅潭地区工业遗产整体设计方案，并提出了保护与更新策略(图 9)。白鹅潭地区在广州的工业发展史上有着重要地位；其除了保留着完整的洋行码头文化遗产外，还是广州工业品牌的“老字号”聚集地。根据《广州市历史文化名城保护规划》，规划范围内列入工业遗产名录的共有 6 处，本次规划在对现状建筑评价后，共划定优秀工业遗产 19 处。[12]

(1) 提出工业遗产利用保护类别区分

在规划中，针对不同类型工业遗产，提出保护利用及保护类别区分，进而提出保留建议和注入功能设想。改造利用类型分为创意产业改造、文化设施改造(如博物馆等)、与购物旅游相结合的商业综合改造、社区公共服务设施、城市开敞空间改造等 5 种(表 4)。其中在城市开放空间上改造明确要求广钢片区规划建设成中央公园，打造成白鹅潭片区重要的开敞空间。[13]

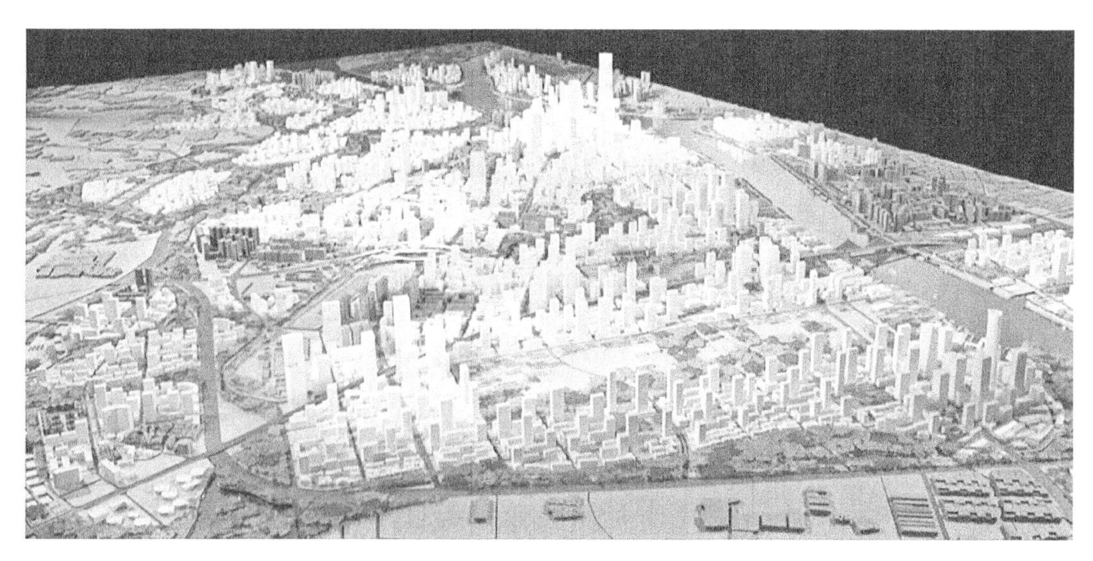

图 9 白鹅潭地区城市设计效果图

表 4 优秀工业历史建筑保护与利用一览表

编号	名称	保留建议	注入功能	历史意义
1	广州铝材厂	最老的建筑及厂区环境	景观改造、茶叶市场配套服务	"前进牌"老字号
2	广州市金源食品厂	特色厂房及大树	景观改造、商业娱乐	广州市唯一以生产南乳、腐乳为主的企业,"雄鸡牌"老字号
3	广州珠江钢琴集团有限公司	北侧厂房 2 栋	展览、设计、娱乐	世界上最大的钢琴厂商,"珠江"牌钢琴老字号(广州钢琴厂)
4	广州柴油机厂	保留厂区整体格局、现状大树及部分特色厂房	创意学校、设计、娱乐	中国柴油机制造业最悠久的企业
5	广州织金彩瓷工艺厂	保留最老的厂房建筑和绿化环境	创意产业改造	"广彩"老字号
6	广州制漆厂	保留老厂房建筑和绿化环境	景观改造、社区服务设施	"电视塔"牌老字号
7	石围塘滨江仓库	铁轨、站台、候车室、龙门吊、滨江仓库等	茶叶博物馆、旅游、休闲	广三铁路遗址、百年车站
8	信义会馆	保留完善周边环境	创意产业改造	现状信义国际创意园
9	金珠江化工厂	近期整饰,远期适当开发	创意产业改造	现状 1850 创意园
10	富民公司储备库	4 栋仓库	旅游服务、商业休闲	特色工业建筑
11	市粮食局杏村仓库	苏式厂房 3 栋	旅游服务、商业休闲	特色工业建筑

（续表）

编号	名称	保留建议	注入功能	历史意义
12	五羊自行车厂厂房	2栋滨江仓库	旅游休闲、商业娱乐	“永久”自行车，“老广”的永久追忆
13	广纸滨水区厂房仓库	沿制浆大道两侧工业建筑、构筑物、广纸水湾、滨江吊机、仓库	滨江公园、社区公共服务、商业配套	广州历史悠久的名牌老厂
14	广钢仓库	部分工业建筑及构筑物等	城市开敞空间、动感产业等	珠三角唯一的钢铁厂
15	广州造船厂	部分历史建筑及临水船坞、码头等	船舶博物馆、开敞空间	珠三角最大的造船厂
16	省百货凤凰仓及周边	民国时期仓库	创意产业改造	本土老牌企业历史，特色工业建筑
17	昌岗油库	油库及构筑物	工业博览馆	特色工业建筑

资料来源：《白鹅潭地区城市设计及控制性规划》

(2) 资源整合，结合工业遗产打造特色环境

充分挖掘白鹅潭地区多样性历史文化资源，并根据文化内涵、历史意义、建筑特色等方面进行分类控制，梳理出对外商贸文化、岭南建筑文化、近代工业文明、地域生态文化、乡土民俗文化五大类。同时白鹅潭地区还保存有大量的码头岸线、铁轨、水湾、船坞等特色工业环境，这些历史形成的地理环境、地形地貌也是珍贵的历史记忆，亦是白鹅潭片区文化复兴的重要元素。规划中应加强细节设计，对特色构筑物、码头岸线加以改造利用，营造浓郁的文化气息，延续历史文脉，制定对特色工业环境的保护与利用措施(图 10)。[14]

(3) 形成分片区导则，指引工业遗产保护

白鹅潭片区被划分为 9 个副地区，每一区各自展现其独特的用地功能分区和开放空间以及城市形态。其中 1 区为商务核心区；2 区为北岸综合生活区；3 区为东岸综合生活区；4 区为旧城综合提升区；5 区为西朗综合枢纽区；6 区为南部综合生活区；7 区为生物医药港；8 区为西部门户综合区；9 区为花地滨水生活区。其中南部综合生活区功能构成为生活居住、商业休闲、中央公园、配套服务等，主要为广钢公园北部和南部街区，以居住功能为主，并包括部分商业和文化功能。同时在白鹅潭地区控规片区规划中，明确要求保留广钢中央公园，这为后面的广钢新城中的工业遗产保护提供了载体(图 11)。[15]

3.4 第三阶段：实地保护与再利用实施

2010 年亚运会之后的广州在继 1990 年代旧城改造之后又迎来第二轮的“旧工厂”改造的高峰，利用老城内大规模的旧工业企业转型外迁实施土地储备以解决地方境务和财政问题。大批近现代工业厂区被划入政府储备出让的清单，这是继 1990 年代近代工业遗产被拆消失以后又一次严峻挑战。广钢作为现代完整的工业遗存也面临消失的危机。为了拯救最后一批可能被保存的工业生产设备、厂房、生产线，遗产调查研究组在修编的《广钢新城控制性详细规划》中提出抢救保护、严格利用的要求。

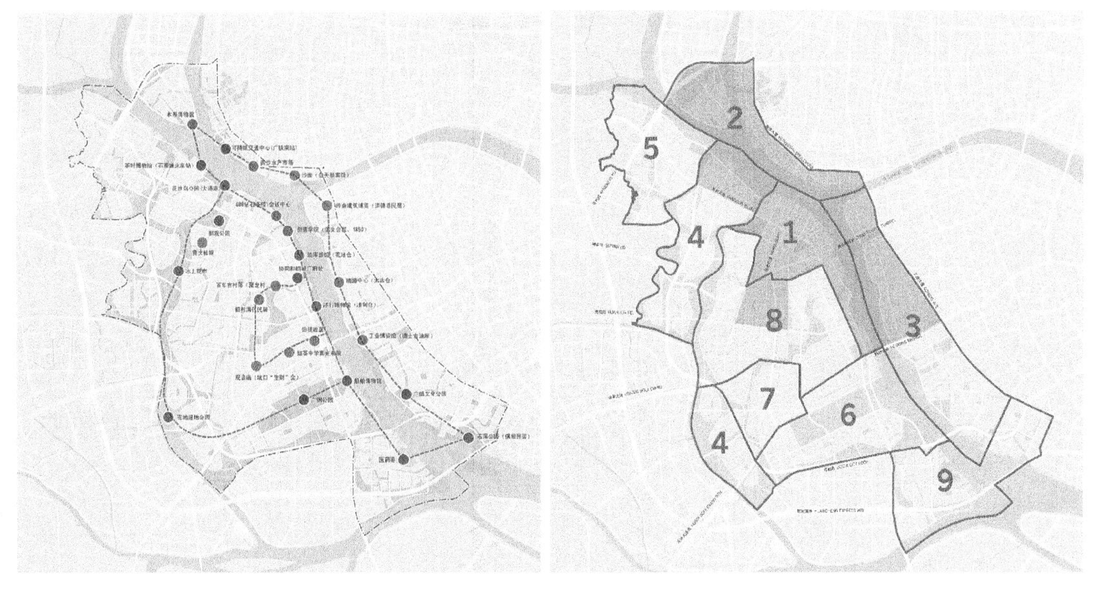

图 10　特色历史文化旅游路线　　　　图 11　白鹅潭控规主导功能图

在 2010 年《白鹅潭地区城市设计及控制性规划》对工业遗产研究的基础上，2013 年完成的《广钢新城控制性详细规划（修编）》开始进入工业遗产保护实施阶段。广钢新城位于花地生态城东南部、珠江西岸，距离白鹅潭商业中心 2.5 千米，进入 21 世纪后广州钢铁厂实现了环保外迁，厂区留下了大量的工业建筑及构筑物。广钢地区是广州近现代工业腾飞发展的重要历史见证者。随着“退二进三”产业发展策略的实施，大量的工业厂房逐渐失去原有用途，且也未被连片开发利用。厂区的工业遗存是历史时期和历史事件的实物载体，也是历史信息最直观的传递者，具有很强的历史价值、经济价值和艺术价值，并传承着时代精神。根据工业建筑遗产评估，确定了 18 类严格保留工业遗产、13 类建议改造工业遗产（图 12、图 13，表 5）。

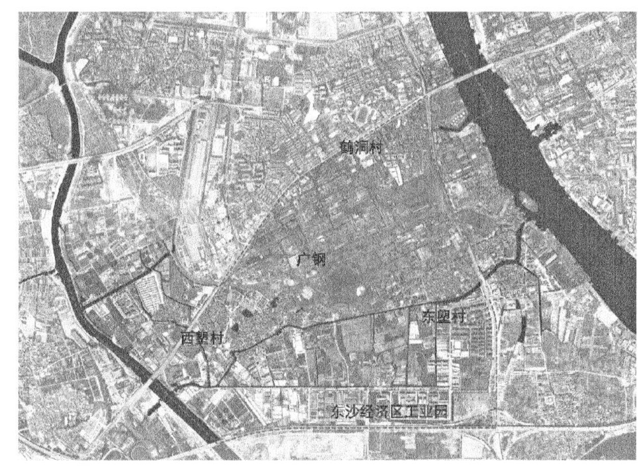

图 12　广钢片区影像图

图 13　2008 年《白鹅潭地区城市设计及控制性规划》广钢片区城市设计

表 5　广钢保留设备清单

类别	序号	名称	地　点	占地面积/平方米	结构类型	遗址类型
严格保留类18个	Y1	原料堆场龙门吊	一号门附近，中和料场东侧、中兴路北侧	5 467	机械型	工业运输设备
	Y2	运输铁路	分布在厂区各处		铁轨型	工业运输设备
	Y3	中和料场配料仓	中和料场内	504	钢筋混凝土	工业建筑物
	Y4	中和料场传送带及铲煤机械	中兴路北侧、中和料场内	16 021	机械型	工业运输设备
	Y5	废钢堆场双梁桥式吊机(现只留下一台)	开拓路东部、中和料场北侧	4 089	钢筋混凝土	建筑及装卸设备
	Y6	矿石转运站	中兴路以南、钢铁大道以北	109	钢筋混凝土	工业建筑物
	Y7	中和料场传送带控制中枢	中兴路北侧、中和料场西端	469	钢筋混凝土	工业建筑物
	Y8	架空运输管廊	炼铁厂、烧结厂和焦化厂各处		砖混结构	工业运输设施
	Y9	水电厂第四净水站冷却池	钢铁大道南侧、开拓路西侧、热电厂西端	989	钢筋混凝土	工业构筑物
	Y10	5 号高炉干法除尘器	炼铁厂内	427	机械型	工业设备
	Y11	3—4 号高炉炉体	炼铁厂内	3 480	钢结构	工业生产设施
	Y12	焦化厂主烟囱	化工北路北侧、一号焦炉南侧	39	砖混结构	辅助生产设备
	Y13	化工北路	焦化厂内	253	道路	厂区交通道路
	Y14	焦炭场双梁桥式吊机	银沙路东侧炼铁厂内	9 100	钢筋混凝土	工业生产设施
	Y15	煤气放散管	转炉车间东侧	7	钢结构	工业设备
	Y16	转炉炼钢车间局部及构架	旧转炉炼钢车间	2 703	钢筋混凝土	工业建筑物
	Y17	5 万立方米煤气柜	燃气公司北部、转炉厂南侧	1 151	钢结构	工业构筑物
	Y18	3 万立方米煤气柜	燃气公司内	932	钢结构	工业设备
建议改造类13个	J1	中和料场机加工间	中和料场内	484	钢筋混凝土	工业建筑物
	J2	中和料场配料仓皮带廊	中和料场内	236	钢筋混凝土	工业运输设备
	J3	热电厂干煤棚	一次料场与中和料场内之间	5 803	钢筋混凝土	工业建筑物
	J4	中和料场火车自动卸料站台	中和料场内	359	钢筋混凝土	工业建筑及设备
	J5	烧结厂主烟囱	开拓路西侧路边	32	钢筋混凝土	构筑物
	J6	出铁场除尘风机房及其烟囱	炼铁厂内	453	钢筋混凝土	工业设备
	J7	铁路调度楼	银沙路西侧、焦炭场以北	172	钢筋混凝土	民用建筑
	J8	2 万立方米煤气柜	燃气公司内	763	钢结构	工业设备
	J9	电炉炼钢厂局部	旧电炉车间、开拓二横路北侧	3 553	钢筋混凝土	工业建筑物
	J10	转炉炼钢厂局部及重力除尘设施	新转炉炼钢车间	5 417	钢结构	工业建筑物
	J11	南部煤场的双梁桥式天车	八号门北侧、原钢渣场南侧	789	机械型	工业运输设备
	J12	无缝钢管厂的双梁桥式天车	无缝厂南侧、东部油库北侧	7 782	机械型	工业运输设备
	J13	轧钢车间	开拓路一横路南侧、鹤洞路北侧	10 508	钢筋混凝土	工业建筑物

(1) 对厂区工业遗产进行整体保护

根据2008年确定的广钢工业遗产公园保护和利用范围和策略，对厂区的整体布局结构框架(包括功能分区结构、空间组织结构、交通运输结构等)以及其中的空间节点、形成元素等进行全面保护。自2014年2月份开始，市政府授权原业主单位广钢对现场一些有场地环境、厂房生产设备运输线路等有价值构筑物实施责任看护、严格管理。这一些举措为未来形成中国首个实物现代工业生产遗址积累下了宝贵的原始物证和经验，为形成首个全面保护再利用的案例奠定了坚实的基础，也为下一阶段研究、设计、开发具体的保护利用提供了法定制度保障和指引(图14—图18)。

图14 原料堆场龙门吊

图15 保留设备

图16 保留运输铁路

图17 广钢高炉设备

图18 厂区保留炼钢设备

(2) 形成工业遗产控制性详细规划导则

根据工业遗产建筑保护名录，制定了广钢新城工业建筑遗产线索保护与再利用规划。对广钢内部各处工业遗产分别制定保护控制图则，划定保护范围、控制建设范围，提出规划建设指引，作为下一阶段项目审批依据。对工业遗产深化细化为严格保留类与建议改造类两类采取严格保护、适度利用、非实物保护等三类保护模式。其中，严格保留类工业遗产"轻易不能动"，就算维修也得严格审批，"修旧如旧"；建议改造类工业遗产可适度改造为城市公园、博物馆、创意产业园或商业综合体；对已消失的重要工业遗产则采取虚拟复原、老设备、档案展示等非实物保护模式。通过对每个保留的工业建构筑物的分析，规划后，厂区将改造成为具有三个层次的景观公园，包括：广钢公园、记忆公园及工业印记空间。规划将广钢的工业遗产集中进行保护与再利用，打造成为后工业景观公园。依托公园，将厂区中各种自然和人工环境要素统一进行规划设计，组织整理成能够为公众提供工业文化体验以及休闲、娱乐、体育运动、科教等多种功能的城市公共文化空间。规划并非对厂区群体建筑及设施进行整体保留，而是根据其在整个炼钢流程中的重要性以及具有的工艺美学价值进行选择性保留(图 19、表 6)。

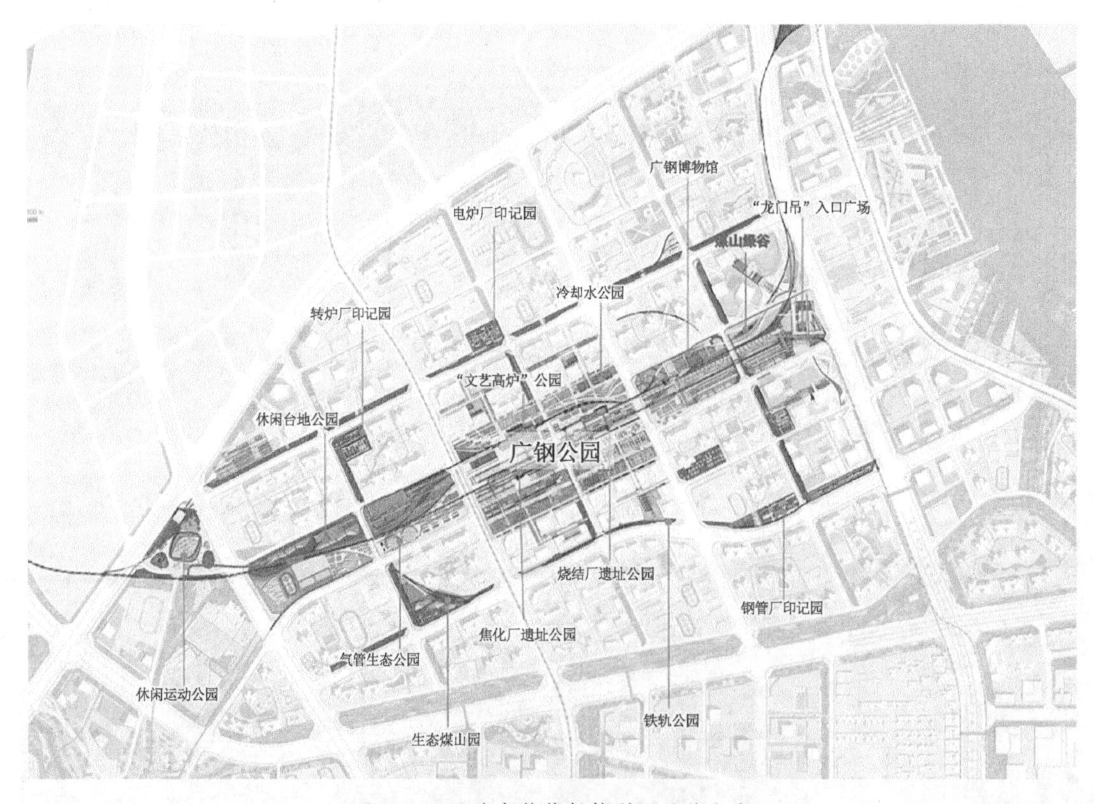

图 19　工业遗产整体保护利用总体方案

表 6　工业遗产地块控制性详细规划导则

工业遗产名称:3、4 号高炉炉体	
工业遗产编号:Y11	
现状基本信息	
所在位置	钢铁大道南侧
占地面积	3 480 平方米
遗产类型	工业生产设备
主要功能	主要生产功能
结构类型	钢结构
体量评估	巨大
保存状况	基本保持原状
周边环境	环境良好
保护要求及改造策略	
保留及改造原因	是广钢的核心,也是制高点,高炉底部有供铁路进入的巨型钢架结构,结构质量良好,与 7 个相同的除尘筒一同具有极高的工业美学价值
改造主要功能	改造再利用
保护类型	严格保留类
改造策略	博物馆模式:主体保留,除污染,使其作为巨型工业展品,向游人展示并传递工业技术文化信息,底部改造为展览馆的一部分,建立电梯走廊,使人们可以登高一览整个广钢公园全景
备注	3、4 号高炉炉台下部结构,高炉北侧的热风炉

4　结语

(1) 工业遗产是人类社会工业文明发展的重要见证[8],也是城市发展的主要记忆和有机组成。广州工业遗产调查、研究的现状和保护实施的例证表明,在中国快速现代化城市建设和更新改造过程中,迫切需要形成全社会共识,政府要率先主导形成各界参与的氛围和规章制度,在当前快速的旧工厂改造主导和土地财政政策压力下,尽快抢救保护快速消失的工业文明遗产成为当务之急。[11] (2)工业遗产是延续城市空间脉络的重要元素,是激发创新实践和建设创新城市的源泉,未来可成为构建地域形态特色和识别性,创造新空间形式艺术改造的重要来源。(3)工业遗产的功能使用应充分尊重依托工业遗产,将其转化为除房地产之外的其他类型新产业形态，可以以市场为主导自发创造低成本就业岗位，以政府为主导推出公益类土地开发供应类别。生态环境和文化继承的因素要逐步融入新的发展主题，以应对城市市场有机更新的有效手段，特别是土地开发、集约节约用地、三旧改造的政策必须要与工业遗产的再利用有机结合起来，防止盲目大拆大建。在产业选择和政策方面给予倾斜和优惠，保障一定的低成本更新，保护一部分设施和开敞空间利用，慎用少用土地单一地采用储备开发房地产的政策。(4)实际保护需要优先确定保护近现代工业遗产的管理机构和及时研究保护利用的技术，发挥其文化教育、科技普及的作用，使得城市的文脉得以延传。

参考文献

[1] 国际工业遗产保护协会.工业遗产之下塔吉尔宪章[Z].下塔吉尔：国际工业遗产保护协会，2003.

[2] 中国工业遗产保护论坛.无锡建议：注重经济高速发展时期的工业遗产保护[Z].无锡：中国工业遗产保护论坛，2006.

[3] 单霁翔.关注新型文化遗产：工业遗产的保护[J].中国文化遗产，2006(4)：10-47.

[4] 刘伯英，李匡.首钢工业区工业遗产资源保护与再利用研究[J].建筑创作，2006(9)：36-51.

[5] 刘伯英，冯钟平.城市工业用地更新与工业遗产保护[M].北京：中国建筑工业出版社，2009.

[6] 杨宏烈.广州工业遗产的保护与利用[J].现代城市，2008(3)：7-10.

[7] 杨希文.广州珠江后航道港口码头工业遗产的保护与再利用[J].铁道科学与工程学报，2011(3)：123-128.

[8] 俞孔坚，方琬丽.中国工业遗产初探[J].建筑学报，2006(8)：12-15.

[9] 张晓辉.广东近代民族工业的发展水平及其特点[J].学术研究，1998(11)：58-62.

[10] 张静辉.工业遗产的保护和利用[J].科技资讯，2008(31)：218-219.

[11] 李建华，王嘉.无锡工业遗产保护与再利用探索[J].城市规划，2007(7)：81-84.

[12] 徐怡芳，王健.广州白鹅潭·洲头嘴地区近代遗产保护措施研究[J].南方建筑，2012(6)：32-35.

[13] 广州市城市规划设计所.珠江后航道两岸近代洋行码头仓库历史遗址保护与利用规划研究[Z].广州：广州市城市规划设计所，2008.

[14] 广州市城市规划设计所，美国SOM有限公司，广州市城市规划编制研究中心.白鹅潭地区城市设计及控制性规划[Z].广州：广州市城市规划设计所，2011.

[15] 广州市城市规划勘测设计研究院.广钢新城控制性详细规划(修编)[Z].广州：广州市城市规划勘测设计研究院，2013.

历史城市内城区建成遗产辨识与价值初探
——以西安明城区回坊片区为例

A preliminary study on the identification and value of the built heritage in the inner city of a historic city: Take the Huifang area of Mingcheng District in Xi'an as an example

李　昊　刘珈毓
Li Hao　Liu Jiayu

摘　要：围绕"真实"与"完整"、"宏大"与"日常"，探讨建成环境类遗产在遗产辨识和价值认知方面存在的问题，解析建成遗产的概念及其内涵。在此基础上，梳理西安明城区回坊片区发展与保护的现实问题，对回坊片区建成遗产的总体特征进行辨识，探讨回坊片区在历史场所的谱系、地方文化的基因和城市意义的彰显等方面的价值属性。
关键词：建成遗产；遗产辨识；价值；回坊片区

Abstract: This paper discusses the problems of built environmental heritage in the identification of heritage and cognition of values by focusing on the heritage of "authenticity" and "integrity", "grand scene" and "daily life", and analyzes the conception and connotation of built heritage as well. On the basis of the above, the paper clarifies the realistic dilemma in the development and protection of the Huifang area of Mingcheng District in Xi'an, identifies the general characteristics of built heritage in Huifang area, and discusses the value attribute of Huifang area in terms of the genealogy of historical place, the genes of local culture, and the manifestation of urban significance.
Key words: Built heritage; Identification of heritage; Value; Huifang area

随着文化遗产内涵与外延的不断拓展，建成环境类文化遗产本体内容与保护观念都发生了深刻的变革。历史城市内城区经历了长期发展与积淀而形成了具有独特文化意义的空间形态及精神内涵，人们开始将文化遗产保护的关注点从遗产本体的保存转向遗产所依存的整体环境。早在2005年《西安宣言》中就曾指出，"古建筑、古遗址与历史区域的重要性在于它们在社会、精神、历史、艺术、审美、自然、科学或其他文化层面的价值，也来自它们与物质的、视觉的、精神的以及其他文化层面的背景环境之间所产生的重要联系"[1]，强调了遗产及其所依存的历史环境之间的重要性。2017年12月，一份名为《关于遗产与民主的德里宣言》(*Delhi*

Declaration on Heritage and Democracy，以下简称《德里宣言》）的重要成果在于印度德里召开的国际古迹遗址理事会第19届大会中通过。《德里宣言》中指出，“遗产是一项基本的人权”，“遗产包括价值体系、信仰、传统和生活方式，以及功能、风俗、仪式和传统知识”[2]。由此，遗产的重要性再一次被提到了全人类福祉的高度，同时这也更加强调了遗产与社区、场所层面的联系，遗产不仅仅记录了一段时期的发展，也与此时人们的生活更加贴近。

1　建成环境类遗产辨识及其价值的现实问题

建成环境类遗产区别于一般的文物，不是孤立的存在。它依托一定的物质空间环境，与社会与人文环境紧密关联，涉及政治、经济、文化和精神等诸多要素，是一种对“变化”的阐释而不是对“静止”的描述。对遗产的“建设性破坏”和“破坏性保护”[3]都是忽略了遗产作为我们了解社会发展状态与变化的直接依据的作用，缺乏对遗产真实性和完整性的辨识，进而造成建成环境类遗产在宏大叙事与日常生活之间的游离。

1.1　貌合神离：遗产的真实与完整

任何一个遗产资源都不是孤立和静态存在的，它们因各种因素相互关联而形成了文化意义上的整体[4]，这就要求对遗产的识别不仅仅限于对其自身的真实性或周边空间范围的完整度，而且要在其形成逻辑、构成特征以及所处社会文化环境的整体背景下进行识别与分析。在遗产研究已成为显学的当下，遗产识别依然存在诸多问题。

首先，早期遗产辨识表现为文物式的“真实”而文脉上的不完整。尤其缺乏对历史环境、文化环境整体性的关照，往往关注最为突出的要素部分，采取了“片段式”的认知方式，对遗产的“真实”关照有加，而对遗产“完整”视而不见。其主要表现为：“见殿不见舍”“见古不见今”“见物不见境”。将遗产本体从共时与历时的语境中抽离出来，致使原本环境中完整的文化信息割裂。失去了真实环境的历史依存，其文化意义日渐被削弱、被遗忘，沦为现代城市中突兀的“文化孤岛”，文物式的“真实”将建成环境类遗产的核心价值旁置。

其次，后期遗产识别表现为风貌式的“完整”而风土上的不真实。街区内众多“风貌建筑”以遗产表现作为符号参照或直接将某一时代的节点风格作为设计元素，刻意营造“历史”，试图塑造“冻结的记忆”；同时真的“风貌区”同样也模糊了真实遗产的辨识度，在“创造历史”与“复制真实”的情境下，大众的生活方式与消费模式在无意之间就被其左右。历史环境是承载原住民日常生活的场所，地方风土逐渐被过度的商业所占据，传统的真实表现被打破，历史街区变成了摹本造就的拟像，看似“完整”而人、物、生活上的真实却早已流逝。两种困境往往相互叠加并贯穿于整个保护过程，致使许多遗产及其环境得不到真正意义上的真实性与完整性的保护，对“真实”与“完整”看似有所考虑实则均有不同程度的缺失。

1.2　厚此薄彼：价值的宏大与日常

人们给遗产的认知与识别赋予了遗产价值，遗产在某种程度上是关于个体或群体信念的再现，与其他信念体系一起共同构成日常生活世界中的一个整体。对于遗产价值的认识，直接影响到遗产保护的实施。《中华人民共和国文物保护法》(简称《文保法》)中将其中一部分文物价值描述为“历史、艺术、科学价值”,《中国文物古迹保护准则》将文物古迹的价值描述为“历史价值、艺术价值、科学价值以及社会价值和文化价值”,在传统价值(历史、艺术、科学)之外,也开始重视其社会、文化等方面的价值并拓展了发展的空间。但在保护实践当中,关注遗产宏大价值的挖掘而忽略其日常价值的情况仍时有发生。

首先表现在对遗产既有价值的认知上存在不足,常常以时间的悠久与否来将保护对象“论资排辈”,多重“古代遗产”而轻“近代遗存”,厚古薄今的价值认识会导致保护行为的局限性,例如仅对具有较高历史与审美艺术价值的遗存进行保护,而忽略近代遗存也同样具有社会与生活方面的意义。其次,将具有突出价值意义的遗存等同于博物馆中陈列的文物一般实施原样保存,将其作为“历史的标本”,往往容易导致只关注外部形态的良好与否,而不关注内部是否也同样具有其他方面的价值。这实则是对遗产价值的片段化认知,忽略了历史所特有的真实叠加的性质,未曾意识到持续发展的、动态的历史才是真正的“活的记忆”。

由于赋予遗产价值的时代变化特点以及对适应性的考虑,遗产价值不只体现在具体事物之中,同时与日常使用、实践与行为及各类传统、信仰都有着紧密的联系,因此对于价值保护的举措也应充分考虑其发展空间。对遗产价值的挖掘并非是宏大愿景的创造,也并非是对“著名”历史的臆想,而是重新审视埋没在“最日常”场所中的暗藏的遗产生活价值,让其得以呈现并存续下去。从文化层面来说,这是对城市地域文化的正确继承与发扬。

2 建成遗产的内涵解读

随着人们对遗产认知的深化,建成遗产(built heritage)的价值观念在时间和空间维度上都呈现出更为广阔的包容性与多样性。作为一种具有历史、社会、经济价值的物质财富,以及城市精神文化呈现载体,建成遗产在真实性、空间范围、社会人文环境等等方面在未来发展过程中都占有举足轻重的地位。

2.1 真实之镜:历史客观性

弗朗索瓦茨·舒艾(Francoise Choay)曾将遗产视为一面镜子:“镜中反映出的影像是当代人对历史记忆的解读。”建成遗产价值认定的核心仍是“真实性”,但建成遗产并不同于文物,后者自出土便可被描述为某一时代的产物。建成遗产从初建起历经世代演变以及无数次人为与自然的变动、整修甚至重建,包容着各式各样的生活与文化样态,其真实性和文物、建筑遗产所要求的真实性多有迥异。建成遗产从始建至今是历史层层叠加的结果,也蕴含着文物建筑所没有的内在叙事逻辑。

2.2 场所之阈:环境完整性

历史城市的魅力来源于其自身不断的新陈代谢和有机生长的过程,在过程中的新旧共存、并置使片区形成了多样化的形态,呈现出不同历史与文化的层层沉淀。将一个区域定义为“遗产”,不应是简单地在其空间范围上设定一个管控的区域,而是要塑造遗产本体与环境共生的整体保护观，将遗存本体与其所根植的文化环境、物质环境关联。其本质上是历史与文化的关联。不同于“形态”上的完整，环境完整要求建成遗产不能随意堆砌与简单累积，不刻意追求空间形态的绝对原样保存，而是要挖掘并延续其依托于文化环境的内在价值，保证各项要素的调和与平衡，把多元的、相互关联的历史文化要素系统地整合起来以体现其历史脉络和时间层积的过程。

2.3 社会之境：生活持续性

建成遗产的演化过程受到经济、文化、社会等各种要素在时间进程中的交织影响，各个时期的社会必然发生动态变迁与发展，更替并持续更替是发展的基本规律。社会互动与联系能够直接影响城市生活与人居质量；同时，城市生活方式、社会组织方式以及人群活动能够极大地影响建成遗产能否维持活态的生活，这直接关系到建成遗产的生存状态。一方面，建成遗产有利于社会的整合，只要为新发展提供适当条件就可以实现功能、社会意义上的拓展，有益于社会的良性、适度的混合，更能避免社会不同阶级的对立；另一方面，建成遗产的特征与其形成的独特的场所精神的维护，也将使城市中多元的、富有特色的地域生活结构得到保护与延续。[5]

2.4 文化之际：多元包容性

文化早已发展成为地区实现效益增长的关键资源，是影响城市精神、地区文化的关键要素。一切行为活动的背后均是特定文化的作用，建成遗产保护与传承的重要意义便是文化的传承与历史记忆的延续。建成遗产对于多元文化的包容有助于形成良好的文化环境。首先，城市文化结构、民风民貌等方面与城市生活一起不断交织叠加，形成了其内在的独特性，这种独特性是实现城市文化认同的精神归属；其次，城市文化在不断发展过程中，对建成遗产也造成了潜移默化的影响；最后，城市文化也为建成遗产所依存的文化环境奠定了基础，为城市文化名片增添了浓墨重彩的一笔。

2.5 发展之源：活态生长性

大卫·罗温索(David Lowenthal)曾这样评价“遗产”：“作为一种活着的驱动力，过去必须不断被更新。”建成遗产中的经济要素是维持发展的重要动力源泉，经由建成遗产效应产生的商品交换、人口流动形成地区特色的交易场所，从产业布局、经济收入上可以为片区发展提供发展的动力，为城市片区带来了充沛的文化商机，使城市片区经济状况持续向好，同时也能为建成遗产营造更好的经济环境。单霁翔先生说过，真正“活”的遗产，是能在现代社会生活中持续发挥作用与价值的。我们不断追求建成遗产的活态化发展，认识到城市的生命周期，接受并理性管控“变化”，让建成遗产能够以一种更好的形态在城市中将文化脉络延续下去。

3 西安回坊片区发展保护的现实问题

回坊位于西安明城区中心，以穆斯林居民为主，是一个保留着地方风土的传统街区。随着城市发展的不断推进,回坊片区的风貌及整体形态正在受到多重因素的冲击:商品经济的发展与旅游业的日渐兴盛,让这个传统的回族社区的矛盾与问题越来越突出;而环境的变迁与往昔生活的更迭正在使人们失去对原有场所环境的历史记忆(图 1)。

图 1 回坊片区 2000—2019 年空间肌理变化

(来源:百度卫星地图)

3.1 日常异化——地域民居逐步消失

1990 年代,在“拆一还一”的更新政策的影响下,居民开始迅速建造以增加其房屋面积[6]。21 世纪初西大街改造工程在一定程度上带动了回坊片区的商业旅游发展,当地居民利用地理位置优势大力发展餐饮、旅游等,用地的紧张使片区整体空间开始竖向发展,片区尺度逐渐发生变化。随着居民住户的变更和人口数量的增加,原有的房屋难以承受高密度的居住,居住环境逐渐恶化;随着居民生活需求的增加,临时构筑物开始侵占院落空间,原有的院落尺度与风貌破坏严重,致使传统院落逐步走向杂乱拼贴的状态。日常生活下的空间形成过程与居民行为方式是人文价值与居住伦理价值的集中展现。在中国传统社会,居住伦理与地缘性、血缘性、家族宗法特征关系最为密切,如今,在城市规划的操控与市场机制的作用结合下对居住人文伦理进行了绝对的控制,人文价值随着城市化进程日渐降低,逼仄的生活环境与快节奏的生活方式加剧了人们精神上的焦虑与陌生,日常生活场景逐渐异化,这对建成遗产的良好存续产生了潜移默化的威胁。

3.2 价值偏差——街区过度商业化

回坊片区身处北院门历史文化街区,旅游业日渐兴盛,商品经济的发展带动了片区经济的发展,同时也带来了更多的问题。回坊片区的主要游览街巷中几乎每一家商铺前面都有 1～2 个自身的临时摊位以承载更多的业态,并且临时摊位与固定店铺的业态重复率较高。商业的招牌形式较为单一,整体色彩繁杂,另外商业环境质量普遍较差,地面脏污、墙面油腻,较大程度地影响了其空间面貌。回坊片区中不断扩张的商业利益需求促使人们为了利益收入开始在片区街巷内肆意拆建、加建与改造。这些改造行为不仅没有考虑原有空间现状并且质量与安全性也得不到保障。另外,回坊片区的生活结构很大程度上取决于穆斯林居民对于宗教信仰的恪守。随着商业化程度的加剧,该片区穆斯林群体共有的价值观念以及教义所要求的礼仪规范等习俗也在潜移默化地发生着改变,个体脱离宗教文化范围的数量逐渐增多,原住民的搬离使社会日常生活主体逐步发生变更。空间在主体替换与标准化生产的双重作用下再一次发生异化，追求最大化利益成为符号消费、文化消费的终极目标。在巨大经济价值面前，对街区的价值认知偏差造成了遗存的历史价值、文化价值、人文价值，场所及其环境的生活价值的重要性日渐被削弱(图 2)。

图 2　回坊片区主要街巷商业化现状

(来源：作者自摄)

4　西安回坊片区建成遗产总体特征的辨识

西安明城区回坊片区作为城市中历史记忆、人文生活同样厚重的地域，其留存至今除了依靠其原有的历史遗存与传统生活之外，还因为整个片区所形成的地域特质往往代表了一个历史片区的精神内核，其所展现的文化意义是需要继承并发扬的。在尊重历史真实性与环境完整性的基础上，重点关注日常生活场景；在梳理历史与当下、社会与物质的本体与生活样态的基础上，辨识其作为建成遗产的特征，明确价值内涵(图 3)。

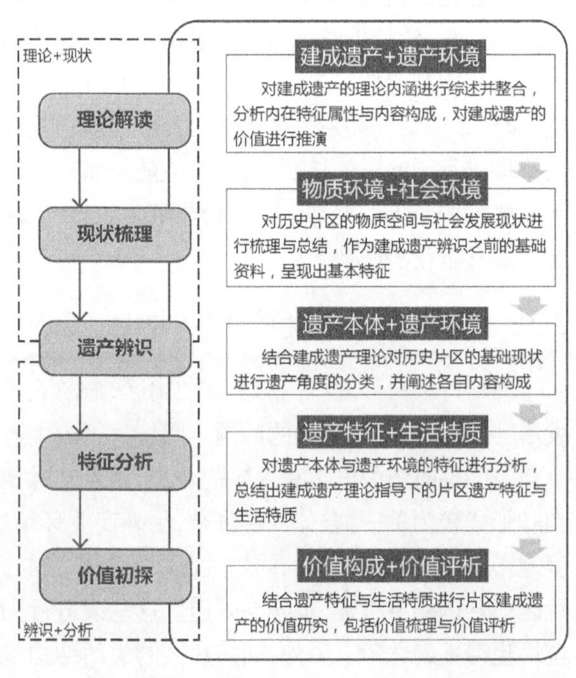

图 3　建成遗产识别与分析框架图

(来源：作者自绘)

4.1　内城核心——历史资源的层积累加区

一个历史片区具有建成遗产环境特征，其首先应与整个城市的环境系统相关联。城市中的历史片区相较于其他新建片区有着更为深厚的历史意义与人文特质，这是一般城市新建片区所没有的。判定历史片区是否具有建成遗产的环境特征，首先应系统、客观地考察该片区与城市历史脉络是否关联。回坊片区位于明城区并靠近其核心腹地，不仅留存了大量的物质遗存与非物质遗存，在西安城市生活中也是生生不息的民族生活社区。这就决定了回坊片区不但记载了过去有关城市片区的历史，并且持续记载着当今片区以及城市的发展，无论是在城市功能还是在城市生活中，都占有重要的地位。如果将回坊片区从明城区的范围搬至其他地区，即使回坊片区的遗产本体要素能完好地保留，但因失去了其赖以生存的历史环境与文化语境，也就将失去文化意义。回坊片区依托于西安明城区，历史的积淀与文化的互通使回坊片区与明城区紧紧地关联起来，回坊片区依靠明城区的生活氛围与历史环境得以生存，同时回坊片区也成为明城区内较为响亮的城市名片。

4.2 格局完整——传统城市的空间意向区

纵观整个明城区，回坊片区以其传统的生活形态与空间结构诉说着回族发展与片区发展的历史，因文化联结成一个连贯、一致的片区，回坊的内在关联性使其展现出的风貌与内城的其他部分有了一定程度的区分。至今回坊片区的街巷空间格局较为完整，仍反映了明清时期的传统空间格局，同时它的整体风貌呈现出较为一致的历史文化特征。回坊片区在回族史上占有着举足轻重的地位，能鲜明地反映历史时期回族生活聚居的特色。回坊片区内仍保留的反映历史风貌的建筑、街巷等是历史原物而不是仿古建造的，不仅包括有形的历史遗存，还包括非物质文化遗产与传统技艺等。历史片区不仅有着真实的文化遗存，同时其维系片区社会组织的结构也形成了该片区的社会凝聚力与在地文化认同。居民对历史片区有着极强的归属感，是因为同样的价值观与生活理念。片区风貌的一致并不代表每一座建筑都属于文物或有记载的历史建筑，但从整体环境的系统性来说，回坊片区风貌的一致性代表着这一地区历史发展脉络的持续，是片区各项系统及要素融合统一的结果。

4.3 民族特征——地方风土的文化富集区

“每一种文化都包含了某种形式的宗教。”[7]回坊片区因居住主体的民族特殊性及宗教因素塑造了地域生活特色，凝结了回族的团结性与认同感。伴随着回族这一群体在回坊片区的发展，其空间生产也经历了不断变化的过程。遗产本体中的回族建筑是伊斯兰文化与中国传统文化互动融合的产物，这使遗产本体同时具有伊斯兰宗教文化与中国传统建筑文化两种特征；再者，回坊片区历史上一次次社会及思想领域的改革也给遗产本体打上了不同的时代烙印，呈现出丰富的动态差异性。世代生活在这一地区的穆斯林居民所形成的价值观念、生活方式以及风俗习惯等，都维系着回坊片区的社会组织结构，使其形成了一种内在的力量。这种力量使生活于之中的居民在行为、精神上具有较强的凝聚力。

4.4 生活持续——活态场所的体验感知区

历史片区在城市中不是一成不变、静态冻结的，它虽比城市其他片区有着更为深厚的历史文化底蕴，但它又同城市的其他片区一样，在城市功能中要持续承担生活的角色。也就是说，一个片区有着深厚文化的同时，必须维持其生活的活力，并且随着城市生活的更迭也要时刻跟紧发展的脚步，考察人们最本质的日常“生活世界”，更加关注人的行为和体验，通过人的参与，将建成遗产同人们生活中的价值和意义联系在一起。回坊片区中的居民、游客及经营者不同的生活表现构成了片区丰富的生活样态。对于游客而言，回坊片区独特的空间文化特征是使其了解一种文化最为直观的方式，他们对一个地方的文化感知是通过公共活动、社会活动等进行的。对穆斯林居民来说，伊斯兰教的教义与教规对其生活方式、行为方式以及价值观念都有着重要的影响作用，他们在宗教信仰之下形成的特定民俗文化（例如饮食文化、丧葬嫁娶风俗及语言）塑造了片区独特的生活状态。回坊片区内呈现出的多元生活样态，构成了建成遗产理论中对生活层面的诠释。

5 西安回坊片区的建成遗产价值初探

通过对西安回坊片区建成遗产总体特征的认知，初步探索其在场所环境、社会生活以及精神文化等方面的价值。

5.1 空间价值：历史场所的谱系

场所在历史中形成，又随着时代的发展而发展。西安回坊片区叠合了多个时代的历史图层，蕴含丰富的历史文化资源与大量珍贵的历史文脉信息。回坊片区整体空间发展至今，呈现出较强的“围寺而居”“依坊而商”的民族社会意识形态在空间上的表达，这在整个明城区甚至西安地区范围内都是较为少见的，是民族聚居的典型空间形态。同时，深厚的历史文化底蕴与当代风貌的复杂性也让回坊片区成了一个矛盾与机遇、历史与当下交叠共生的典型区域。其中清真寺作为回坊片区中民族社会结构的核心，其空间组织、形制形态均能够呈现出中国传统文化与民族宗教文化的融合，有着较高的艺术、审美以及历史价值；同时它也是穆斯林居民的信仰中心，是日常举行宗教活动的场所。对该场所的持续使用是对地域生活、群体观念与信仰的尊重与理解。

5.2 社会价值：地方文化的基因

多元的生活方式与多样的生活场所赋予了回坊片区独特的生活意义，使其成为明城区内乃至整个西安城市内都较为独特的片区。生活在回坊片区的人们延续的生活方式及对周围环境的改造和创作所蕴含的“场所精神”让我们能透过片区的整体生活氛围和物质遗存切实感受到生活的历史气息。以清真寺为场所中心的回坊片区，形成了特殊的“街—巷—寺—坊”空间结构：一方面，在片区中穆斯林居民信奉伊斯兰教而形成的各种生活方式，所传达的社会生活价值凝结在生活场所之中，也因宗教的认同感而产生个体的归属感；另一方面，民族亲缘、地缘情感的维系，也使这里的居民形成强烈的地域和家园意识。回坊片区因民族与宗教因素而形成的强大的凝聚力促使回坊居民与片区、外来游客与片区之间形成了紧密的利益联系，这使回坊片区的资源与利益能良好地共享。同时，片区生活的繁荣为回坊空间的发展带来了持续的动力，良好的片区管理机制保证了片区的活力有序发展，也为片区增添了特色。

5.3 精神价值：城市意义的彰显

回坊是回族民众构成的社区，清真寺始终保持着对居民的强大吸引力。片区独特的宗教文化是其空间形成与发展的主要动力，从最开始的“小聚集、大分散”到最终形成的大的民族聚居区,从回坊的历史演变与空间发展即可看出回坊片区因宗教而集聚形成的功能文化区的特征,同时也集中体现了不同民族文化的特征。回坊片区承载了大量历史文化资源,其中丰富多元的历史信息赋予了回坊片区空间一定的历史感,融合了精神、审美以及艺术上的意义,而居住于其中的居民,将居民自身对这片区域的情感与空间的历史气息相融合,找到了文化上的认同与归属。文化是一种传承的、有着时间属性的概念,它延续着一个地区的在地特色,也是这个地区永续发展的前提和基础。对于上述要素的系统认知与尊重、认同,是维持回坊片区历史遗存保护、生活环境良性发展的必要价值认知前提,同时也是延续城市文化多样性的客观需求、对城市多元文化精神的阐释。

6 结语

回归以价值为核心的保护,从只强调实体静态保护上升至对当代价值与文化意义的关注,人们对建成遗产保护的核心要义得以清晰,即从"保护什么"向"为何保护"再到"为谁保护"。正如大卫・罗温索所说:"每个遗迹都是一份证言,其不仅属于它的创始人,也属于后继者,它不仅承载了历史之精神,也展示了今日之思想。"建成遗产是历史的记录与生活的见证,遗存不应该是已经"死去"的标本,相反它是一个生命有机体,它始终承载着人们日常生活。历史片区作为城市文化底蕴最为深厚的地方,同其他片区一样随社会发展而动态演进,在延续历史文脉的同时享有现代生活。需要强化对在地日常生活的尊重与维护,让外来体验者能够从片区的日常生活氛围中感受到其独特的文化内涵和人文精神,这有益于构建历史环境与当下生活之间的良好共生关系。建成遗产在构成城市中范围最广、数量最多,受城市快速发展影响也最直接,不同文化之间的冲突与矛盾也最为彰显,它"既是先人活动的遗存,又是今人生活的空间"。实现城市历史资源的保护利用与空间场所的活化更新,保持其与城市地区经济、文化、生态、社会的持续健康发展仍需要一个漫长的研究实践探索过程。

参考文献

[1] 国际古迹遗址理事会.西安宣言(2005)[M]//联合国教科文组织世界遗产中心,国际古迹遗址理事会,国际文物保护与修复研究中心,等.国际文化遗产保护文件选编.北京:文物出版社,2007.

[2] ICOMOS.德里宣言:遗产与民主[EB/OL].(2018-01-08)[2020-04-28].http://www.icomoschina.org.cn/content/details90_2522.html.

[3] 李昊,刘珈毓.厚度+活态:从历史静物走向共生遗产:西安明城区保护规划策略[J].科技导报,2019,37(8):61-67.

[4] 邵甬,陈欢,胡力骏.基于地域文化的城乡文化遗产识别与特征解析:以浙江嘉兴市域遗产保护为例[J].建筑遗产,2019(3):80-89.

[5] 张松.历史城市保护学导论:文化遗产和历史环境保护的一种整体性方法[M].上海:上海科学技术出版社,2001.

[6] 翟斌庆.中国历史城市的更新与社会资本[M].北京:中国建筑工业出版社,2014.

[7] [美]菲利普・巴格比.文化:历史的投影[M].夏克,等译.上海:上海人民出版社,1987.

“城市愿景 1910/2010”展览
——百年前/后的城市设计学科发展

“CITY VISIONS 1910/2010” Exhibition：Urban design discipline’s development at the beginning & end of 100 years

易　鑫
Yi Xin

摘　要：本文围绕2010年举办的“城市愿景1910/2010”展览，对百年前后的城市设计学科发展情况进行了总结。文章介绍了1910年柏林举办了一次对于城市设计学科发展具有重要意义的“一般性城市设计展”，并结合1908年大柏林城市设计竞赛的成果，对现代城市设计的早期发展进行了梳理。1910年代城市设计学科的专业主题包括6个方面：塑造大规划，城市中心地区的纪念性，寻找替代高密度内城地区的发展方案，发展新的花园式郊区，绿带、绿楔与公园体系，城市区域内部的机动性问题；2010年代城市设计学科的专业主题包括8个方面：城市中心成为整个城市区域的窗口、传统工人街区的改造转型、社会住宅区的复兴、城市废弃地的再利用、克服郊区蔓延和景观破碎化的措施、发展绿色城市、应对气候挑战和发展可持续的交通模式、对战略性规划的推崇。
关键词：“城市愿景1910/2010”展览；城市设计学科发展；柏林

Abstract：This article revolves around the “City Visions 1910/2010” exhibition held in 2010，and summarizes the development of urban design discipline during a hundred years. The article introduces a “general urban design exhibition” held in Berlin in 1910，which is of great significance to the development of urban design discipline，and combined with the results of the Greater Berlin Urban Design Competition in 1908，it combs the early development of modern urban designs. The professional themes of urban design discipline in the 1910s included six aspects：shaping the grand plan；the monumentality of the urban center；looking for alternative development plans for high-density inner city areas；developing new garden suburbs；green belts，green wedges，and park systems；mobility issues within the urban area. The professional themes of urban design discipline in the 2010s include 8 aspects：the urban center being the window of the entire urban area；the transformation of traditional worker blocks；the regeneration of social housing areas；and the reuse of urban wasteland；measures to overcome suburban sprawl and landscape fragmentation；the development of green cities；responding to climate challenges and developing sustainable transportation modes；and promotion of strategic planning.
Key word：“City Visions 1910/2010” exhibition；Urban Design Discipline Development；Berlin

2010年10月,在德国"联邦交通、建筑与城市发展部"的支持下,柏林工业大学建筑论坛举办了"城市愿景1910/2010"(CITY VISIONS 1910/2010)展览。本次展览以纪念100年前在柏林举办的一次对世界城市设计学科发展具有重要意义的"一般性城市设计展"为契机,在回顾展览的同时,对今天的城市设计学科发展进行了展望(图1)。[1]

图1 "城市愿景1910/2010"展览海报*

1.1 历史缘起与条件

1910年5月,柏林举办了一次对于城市设计学科发展具有重要意义的"一般性城市设计展"(Allgemeine Städtebau-Ausstellung)。展览由柏林视觉艺术高等学校(Hochschule für die bildenden Künste)即今天的柏林艺术大学主办。[2]

作为在欧洲相对年轻的城市,经过德国1871年统一之后的发展,柏林一跃成为与伦敦、巴黎、维也纳、芝加哥、纽约与波士顿等并列的世界级大都市。本次展览为欧美当时重要的城市设计及城市规划的学者提供了重要的交流平台,世界上最具影响力的城市规划师和城市设计师如英国的雷蒙德·昂温(Raymond Unwin)、法国的尤金·海纳德(Eugene Hénard)、美国的丹尼尔·伯纳姆(Daniel Burnham)、德国的海尔曼·延森(Hermann Jansen)在展览期间进行了广泛的交流,共同开启了对于未来城市进行思考和规划的新时代。

柏林于1908年发起了大柏林设计竞赛(Wettbewerb Gross-Berlin),于1908年10月15日正式向外界公布,参赛作品提交的截止时间为1909年12月15日。竞赛组委会后来决定以本次竞赛为契机,组织一次"能够体现整个城市设计领域"发展的国际性展览,"包括建筑艺术、经济和政治等方面内容，推动塑造整个大柏林的发展和升级”[1]（图 2）。

除了在柏林举办以外，本次展览的成果还分别在 1910 年 10 月于伦敦的“城市规划会议暨展览”（由英国皇家建筑师学会主办）和 1913 年于比利时根特的世界博览会再次展出。本次专业性展览共吸引了 65 000 名观众，在学术界和当地社会取得了广泛而积极的反响。

图 2　大柏林设计竞赛获奖方案及说明书的封面*

1.2　现代城市设计学科的建立

一战之前的德国是世界上城市设计学科发展的中心，在首都柏林所进行的城市设计项目是工业社会应对城市问题最重要的试验之一（图 3）。其核心的目标是希望为处于工业快速增长期，同时面临混乱局面的大城市提供秩序并发展出类型化的处理模式。展览组织者将当时柏林城市发展所面临的问题总结成以下主题：社区的混乱状态，大城市交通发展不完善，工人阶层的居住环境恶劣，缺乏绿地空间并由此造成休闲设施不足，公共空间条件不良，特别是空间形态水准较差。

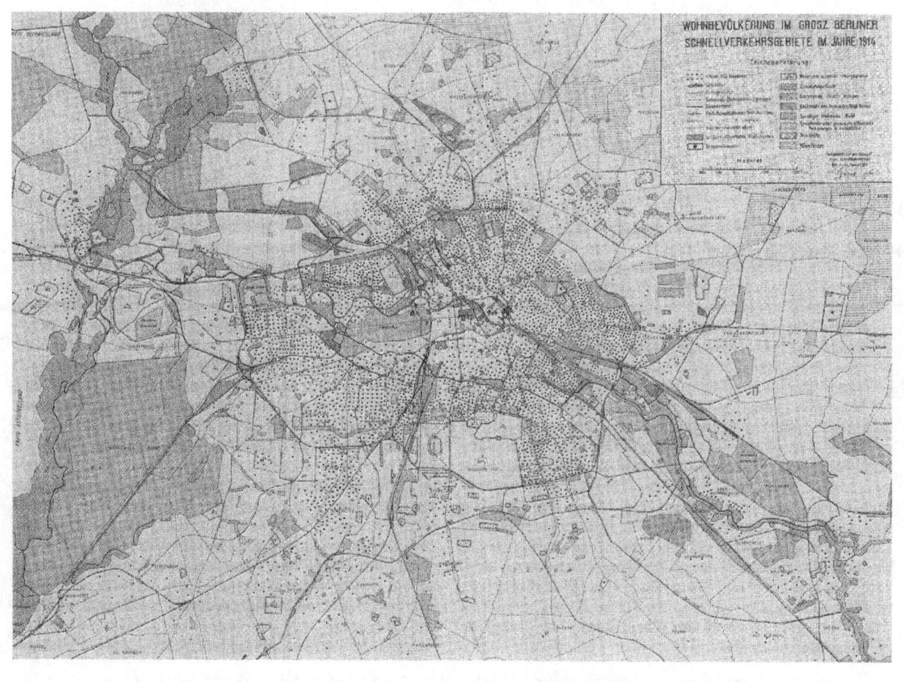

图 3　1914 年柏林的人口、基础设施以及绿地分布情况*

德语中的"Städtebau"(城市设计)概念产生于 19 世纪末,在世界范围内产生了广泛的影响。[3] 当时的学者认为城市设计工作的核心是关于规划与形态问题相结合的方面,致力于实现城市的建造过程与最终的成果统一,进而对规划、实施和建成空间本身进行反思。

1910 年前后是城市设计学科的繁荣时期,在这一时期,受到多学科的影响,再加上之前长期实践经验的积累和国际性的知识交流,相关的学者开始对城市发展进行系统的构想。与 19 世纪城市设计领域主要受到工程师影响、主要关注城市的功能与技术侧面不同,城市设计师开始引入城市的空间形态维度,并以空间美学为催化剂,推动了对城市设计领域一系列复杂问题的系统解决。

对于艺术与美学的关注,使得城市设计开始重视其跨专业的整体性要求,将技术、卫生和经济等问题结合进来。在 1908 年竞赛公布的文件中要求:规划设计成果"既要服务于交通,也要能够满足在美学、人民健康(卫生)和经济方面的要求"。

虽然 20 世纪早期的城市设计面对着包括交通、住宅、卫生与绿地空间等急迫的问题,但是美学问题成为统筹解决以上问题的切入点,帮助各个领域的专业人员、政府官员、开发商和市民勾勒出一幅明确的城市意象,虽然后来的学者对蓝图式规划方式进行了批判,但是通过图像引导共同的规划目标构想,对于城市设计学科以及城市发展有着深远的意义(图 4、图 5)。

图 4　大柏林设计竞赛:新歌剧广场设计鸟瞰图(Eberstadt, Möhring und Petersen)*

一战之前的欧美城市设计开展了广泛的交流,在形成以各自国家实践为特色的学派的同时,也通过成立相关的学术组织、学术刊物、国际会议,大大促进了城市设计思想和知识的国际性传播,其影响力也超越了单纯的文化圈,渗透到社会主流的内部。各个国家举办了具全国性影响力的学术会议和展览,1903 年德国在德累斯顿举办的第 1 届城市设计展、1907 年英国在伦敦举办的国家田园城市城镇规划会议和 1909 年美国于华盛顿举办的首届国家规划会议,都举办了系列展览。

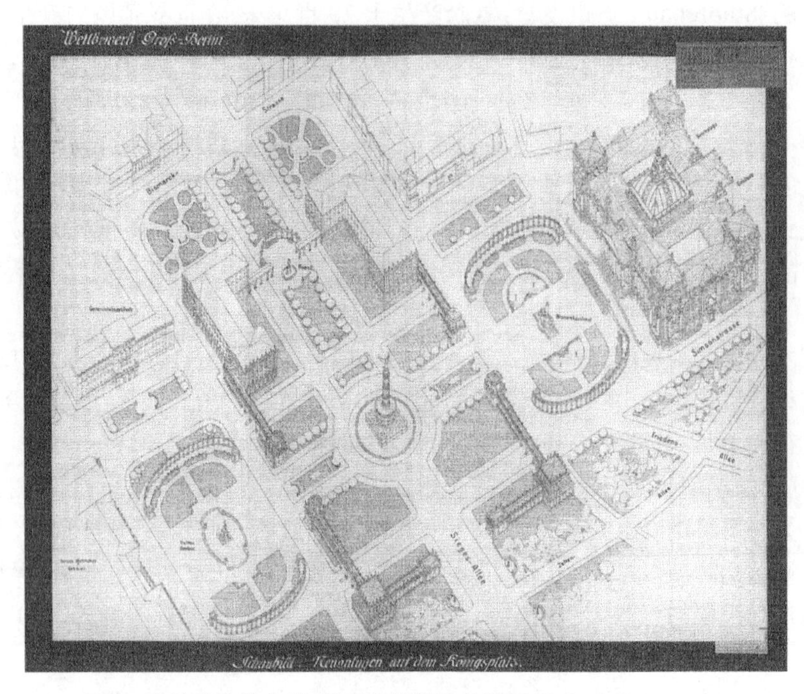

图 5　大柏林设计竞赛：国王广场设计轴测图(Brix & Genzmer)*

1.3　国际城市设计展的组织

与伦敦在 19 世纪的蔓延式城市化相比，柏林的城市扩张通过铁路交通线等基础设施的引导，城市建设区相对紧凑。但是 19 世纪的柏林与其周边的社区在规划方面缺乏相互协调，且城市与社区之间的竞争，再加上大型私人开发商的参与，使得城市发展面临复杂的局面。城市一方面通过私人开发商，在城市外部，特别是西部和南部兴建了很多具有特色的中产阶层居住区，另一方面也面临严重的社会分化，尤其是严峻的居住压力问题，超过 100 万人口的工人阶层的居住条件非常差，每个有供暖房间的居住人口从 3 人到 13 人不等！由于受到政府在建筑限高和防火等方面的约束，私人开发商兴建了大批供出租的多层住宅楼，但由于其过高的密度和狭窄阴暗的内院的特点，被后人戏谑地称为"租赁的营房"(Mietskaserne)。这些问题也暴露出，由于当时奉行的自由放任政策，国家的不作为造成城市设计成为加剧社会分化的工具，引起了当时学者的普遍反思(图 6)。

除了对于居住问题本身的关注以外，为了使柏林市中心具有世界级城市的空间纪念性，城市有必要通过加强周边城镇的服务职能来平衡城市中心区以美学为重点的发展要求。再加上 1900 年前后，城市设计领域进一步引入了历史和自然环境保护问题①，这些保护工作进一步对现代大城市地区的无序蔓延问题提出广泛的约束性要求(图 7、图 8)。

① 德国 1904 年在德累斯顿成立"家园保护协会"(Bund Heimatschutz)，1906 年柏林又成立了"森林保护联合会"(Waldschutzvereins)。

图 6 “租赁的营房”实例*

图 7 普鲁士市中心宫殿群全景(今博物馆岛)*

图 8 柏林的巴伐利亚广场鸟瞰图*

1905 年，伊曼纽尔·海曼(Emanuel Heimann)与西奥多·格罗克(Theodor Goecke)等人在柏林建筑师协会提出制定“统一的大柏林地区建设规划”，该想法于 1906 年获得批准。1907 年有人提议为此举办一次国际性设计竞赛，目标是“通过深思熟虑，并结合艺术性精神的协调……为将柏林发展成 20 世纪在技术和美学上相统一、有影响的大城市做出准备”[1]。除了柏林之外，周边的中小城市社区(Charlottenburg，Schöneberg，Rixdorf，Wilmersdorf，Lichtenberg，Spandau，Potsdam)也积极支持和参与。截至 1909 年 12 月 15 日，组委会共收到 27 份参赛作品。

大会最终评出两个一等奖方案[Hermann Jansen 的方案(图 9)与 Joseph Brix、Felix

图 9 竞赛一等奖方案：Hermann Jansen *

Genzmer、Hochbahngesellschaft 的方案],一个三等奖方案(Brono Möhrig、Rudolf Eberstadt),一个四等奖方案(Otto Blum、Havestadt、Contag),未分配五等奖。另外购进了四个方案[Albert Sprickerhof 的 Alber Gessner(图 10),及 Fritz Kritzler、Hermann Jansen 另外所做的局部地区规划设计方案]。

图 10 大柏林设计竞赛:总体规划鸟瞰图(Alber Gessner)*

这些方案在之后被根据不同的策略和专业侧重点进行研究和发展,用于制定最终的大柏林建设规划,其重点包括:

■ 实现交通系统的整合,在郊区调整并增加轨道交通站点,引导居民点体系的建设。(图 11)

■ 绿地空间的整合,在城市中心和郊区的开放空间内兴建林荫大道,同时规划公园和体育设施网络。

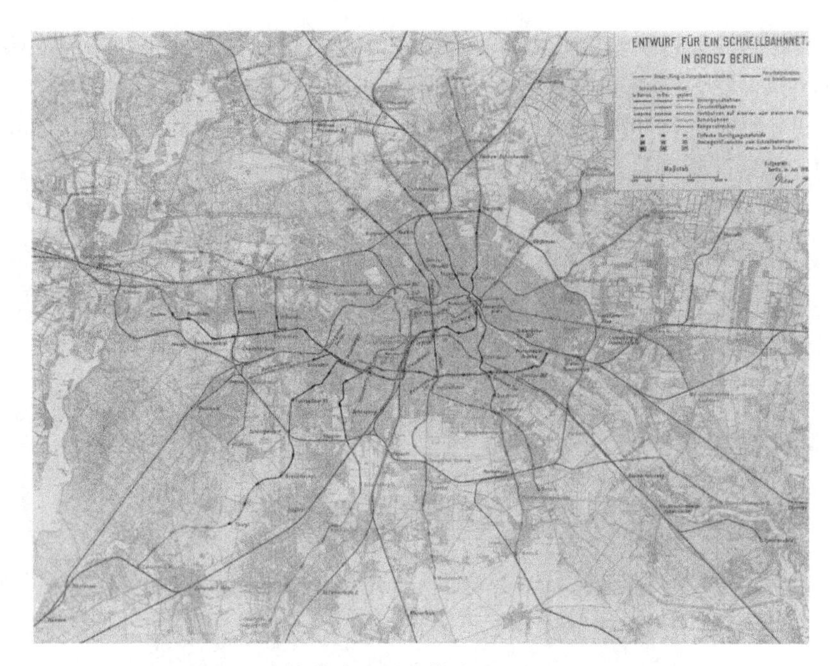

图 11　大柏林地区区域轨道网络规划(1916)*

■ 塑造统一连贯的城市空间形象。

■ 实现交通系统与绿地空间在区域尺度上的关联，在道路、统一的建筑群与绿地开放空间之间建立有机联系。此外还建议将城市外围的轨道交通发展与绿楔等结构结合起来发展(图12、图13)。

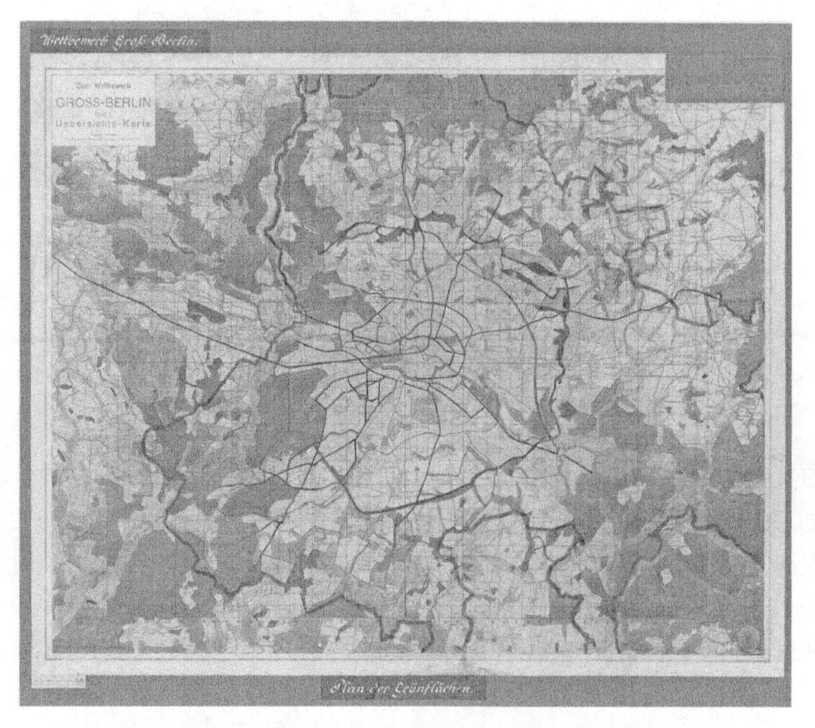

图 12　大柏林设计竞赛：绿地空间总体规划图(Brix & Genzmer)*

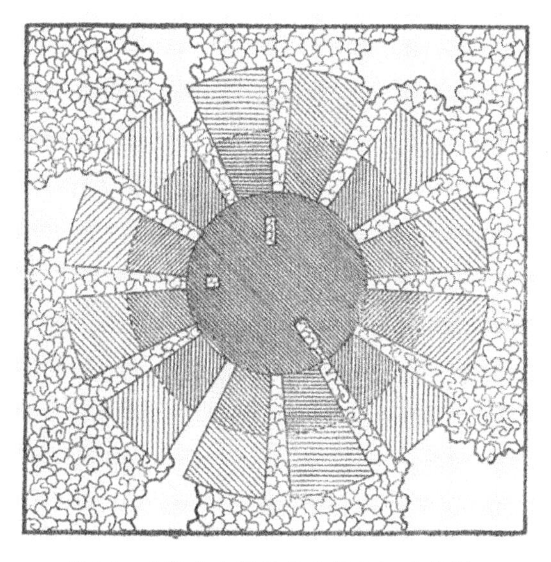

图 13　大城市空间绿地空间组织示意图*

1.4　1910年代城市设计学科的专业主题

1910年的展览对来自世界各国的参展内容进行了梳理，共总结出以下六个方面的主题：

■ 塑造大规划：面对现代大城市发展的复杂要求，城市设计在从美学角度出发的同时，需要在更大的尺度上结合急迫的居住、交通、绿地空间等问题，构想出综合性的城市发展空间（图14）。

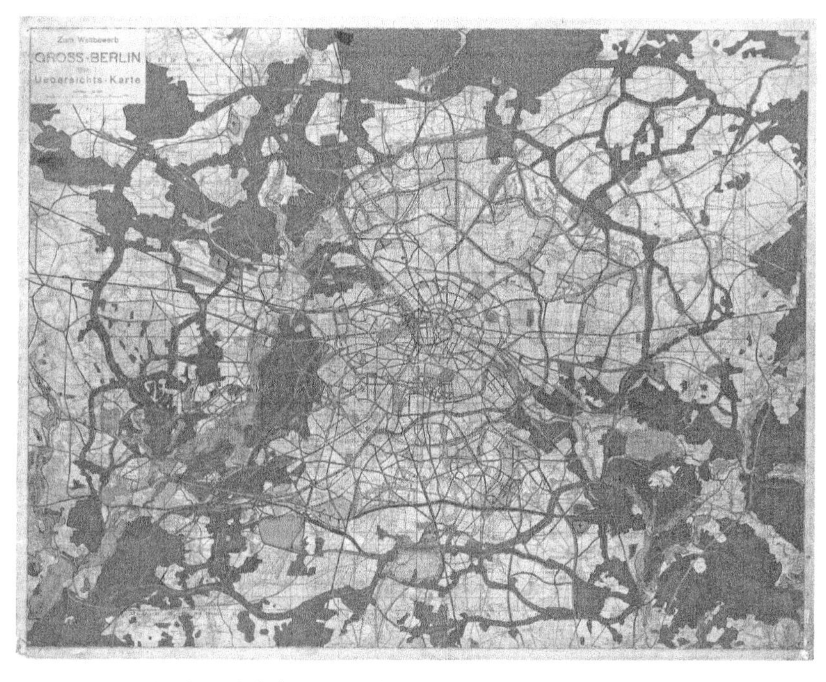

图 14　大柏林设计竞赛总体规划图（Léon Jaussely und Charles Nicod）*

■ 城市中心地区的纪念性：由于城市在区域层面的职能分化，及政治、经济、社会和文化等方面的多样性要求，历史性城市中心发展的重点在于发展包括宗教、宫殿、文化等设施，塑造以展示为重点、具有鲜明特色的纪念性中心(图 15、图 16)。

图 15　大柏林设计竞赛：国王酒馆大街拓宽设计方案图(Brix & Genzmer)*

图 16　大柏林设计竞赛：国王大街设计方案图(Brix & Genzmer)*

■ 寻找替代高密度内城地区的发展方案：各国专家都开始寻求在城市的外围地区发展具有更好的绿地和卫生条件的居住方式。

■ 发展新的花园式郊区：霍华德(Howard)提出的“发展独立于大城市以外的中小城市”的田园城市思想获得了广泛的赞同，并切实开展了在列契沃斯(Letchworth)和韦林(Welwyn)

等地的初步实践(图 17)。此外还出现了大量从属于大城市的以居住功能为主的郊区居民点(图 18、图 19)。

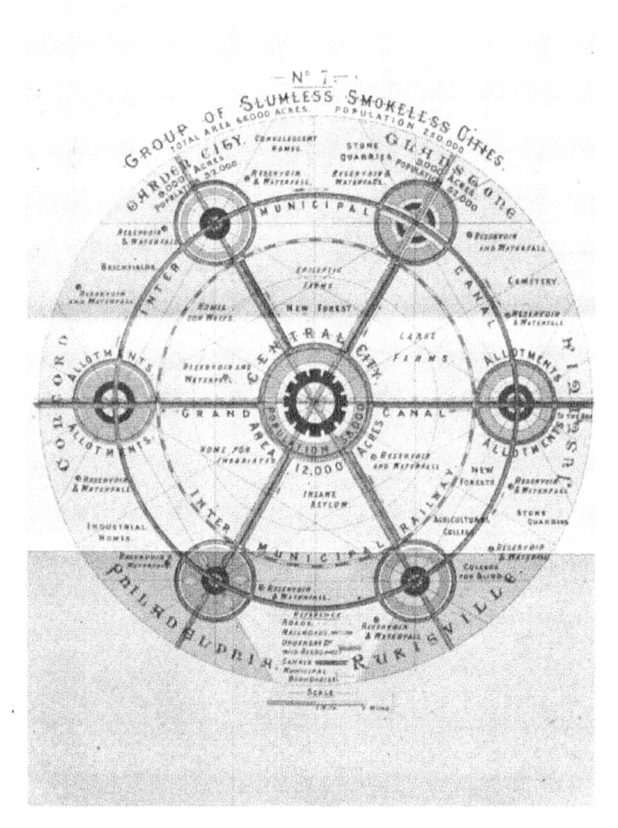

图 17　霍华德的田园城市概念示意图*

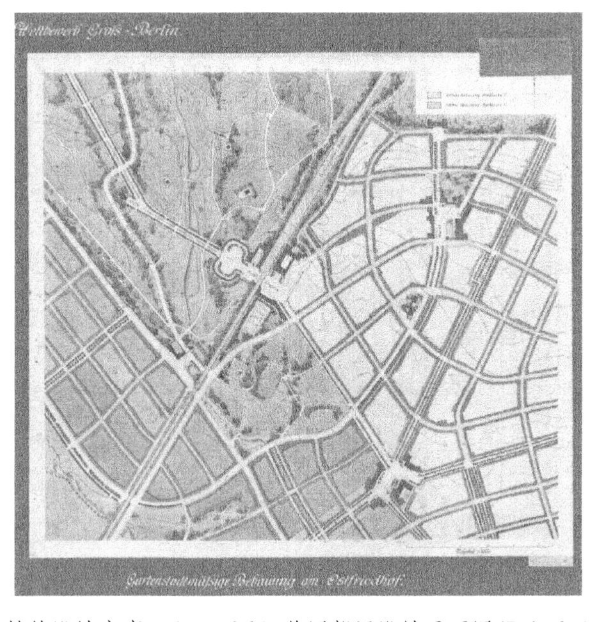

图 18　大柏林设计竞赛:Ahrensfelde 花园郊区设计平面图(Brix & Genzmer)*

图 19　大柏林设计竞赛：Ahrensfelde 地区集市广场设计鸟瞰图(Brix & Genzmer)*

图 20　高架铁道设计建议(1909)*

■ 绿带、绿楔与公园体系：根据大城市地区内部空间要素的多样性特点，追求在城市内部发展绿地空间，以及区域尺度的自然环境保护，推动城市与区域景观的发展。奥姆斯特德兄弟在美国提出了发展全面的“公园体系”的设想。

■ 城市区域内部的机动性问题：现代大城市在区域尺度上的空间重构离不开交通机动性的支持，由于当时的机动性还是以轨道交通为主，交通工程师除了对区域铁路网络和城市交通设施的规划建设进行研究以外，还提出了架空轻轨系统等新的交通方式(图 20、图 21)。

1.5　对柏林城市建设的影响

从柏林竞赛的内容可以看出，20 世纪初，以美学问题为切入点，城市设计学科开始致力于将居住问题、城市交通、绿地空间发展整合起来，以艺术性原则提出城市发展的整体性视角。竞赛的成果反映了当时城市规划所关注的一系列问题，包括对城市地区的总体规划与控制、与内城高密度地区相区别的其他城市建设方式、花园城市或花园郊区、城市内部及外围的绿地与开放空间，以及城市区域内部的交通机动性等(图 22、图 23)。

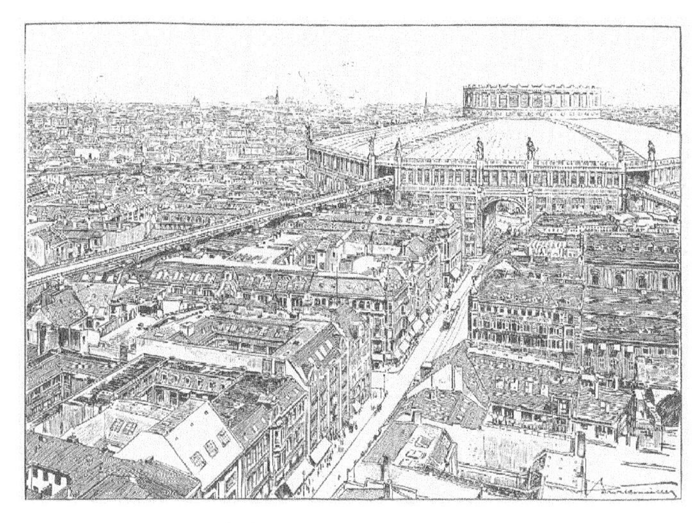

图 21　服务高架铁道的中央车站设计建议(1909)*

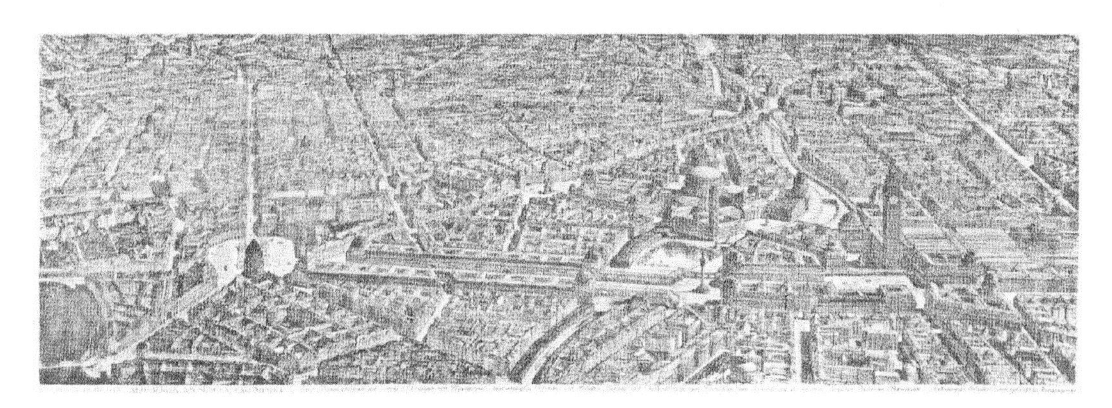

图 22　大柏林设计竞赛:南部中央车站地区设计鸟瞰图(Havestadt & Contag, Schmitz und Blum)*

图 23　大柏林设计竞赛:滨河街道透视图(Hermann Jansen)*

根据西奥多·格罗克和卡米罗·西特(Camillo Sitte)的观点，城市设计学科致力于"将各种技术和造型艺术统一在一起"。城市设计需要对建设活动进行协调和调控，"在现实中，技术、功能与美学是作为整体得到实现的。建筑物与外部空间的布局，不仅仅只是被居民所使用，同时也能够被人们(有意或无意地)体验到，并由此产生美学方面的效果。同样，对于建筑物与外部空间的布局、确定准备建设或不准备建设用地的规划，也不只是技术上的考虑，而恰恰是依据规划者在美学方面的判断(所希望的和所不希望的)所决定的，因此可以说"……'实效的'和'美学的'任务是不可分割的"[5]。

在本次竞赛的影响下，德国部分地区出现了专门的公共机构，尝试处理与区域规划有关的问题。1912 年柏林成立了德国最早的具有区域规划性质的机构——"大柏林地区协作联合会"(Zweckverband Groß-Berlin)。

1920 年，普鲁士州议会做出决定，确定了大柏林地区的行政边界。至此，柏林的范围大为拓展，由原先的 66 平方千米扩大到 878 平方千米，人口由 190 万扩大到近 390 万人。这一决定，让柏林原先过于拥挤的城市中心得到逐步疏散，同时也推动了城市的功能结构调整。由此柏林成为当时欧洲继伦敦之后土地面积第二大的城市、继伦敦和巴黎之后人口第三多的城市。

后来的工业建设项目得以更多地在城市外围地区选址，同时原来的城市中心地区随着普鲁士王宫的外迁而转化为服务业中心。柏林的城市中心地区转而得以逐渐发展出具有重要展示职能的纪念性城市空间。

1.6 2010 年代城市设计学科的专业主题

2010 年 10 月 15 日至 12 月 20 日期间，柏林工业大学建筑论坛举办了第一次"城市愿景 1910/2010"展览。展览以纪念 100 年前在柏林举办的一次对世界城市设计学科发展具有重要意义的"一般性城市设计展"(Allgemeine Städtebau-Ausstellung)为契机，在回顾 1910 年展览的同时，对今天世界上城市设计的发展进行了展望。该项目由柏林工业大学、魏玛包豪斯大学、多特蒙德工业大学和法兰克福德国建筑博物馆的学者共同策划。在展览团队的设想中，"城市愿景 1910/2010"展览将和 1910 年的展览一样，使广泛的公众更加直观地了解城市设计问题，从而能够让公众理解和支持城市设计。

展览对四个国际性大都会——柏林、巴黎、伦敦、芝加哥——在百年前后的城市建设进行了比较，以反思的方式对城市发展、城市建设的特征与内涵进行了批判性的考察。

到了 2010 年，当代西方的城市设计任务主要包括八个方向：

■ 城市中心成为整个城市区域的窗口：大型城市区域的中心地区正经历着让人难以忽视的复兴。自 1980 年代以来，以前长期的去中心化趋势减弱，取而代之的是一个再中心化的转变过程。除了吸引了全世界的游客以外，城市中心也在城市区域参与国际竞争中发挥了主要的宣传作用，在吸引公司和各种创新性群体落户等方面发挥了关键作用(图 24)。

■ 传统工人街区的改造转型：在去工业化过程中，此类地区的许多居民失去了工作岗位。如今居住在里面的居民很多都是背景复杂的移民，而且经济能力较弱，不过他们已不再是以前的产业工人。以前人们一直忽视这些街区在强化城市中心区方面的历史价值，然而恰恰是这些历史价值能够帮助这些位于市中心的内城工人聚居区焕发出最大的活力。

图 24　2012 年重建之后的"巴黎广场",附近为勃兰登堡门*

■ 社会住宅区的复兴:在欧洲城市的区域中,有很大一部分人口住在作为社会福利住宅的大型居住区里面,这些居住区主要是在 1960—1980 年代开发建设的。这样的大型居住区因为暴露出各种城市设计方面的缺陷,现在往往声名不佳。城市设计致力于引入小尺度的低层建筑群,从而与那些住宅楼形成相互平衡关系。此外,细分的地块结构有助于加强房屋的产权关系,并更好地促进社会阶层混合(图 25)。

图 25　柏林的大型居住区得到改造*

■ 城市废弃地的再利用:这些废弃地包括昔日的工业、商业、港口、机场和军事用地等多种类型。这些往往位于内城的大片土地为城市设计工作提供了重要机遇,这些地区的开发具有相当大的潜力，不过也存在风险。这些地区吸引了大量投资机会，它们被用于安排各种功能混合的街区，比如开发内城住宅、新的城市公园以及其他一些特殊功能，这就给各种混合使用提供了很好的发展条件(图 26)。

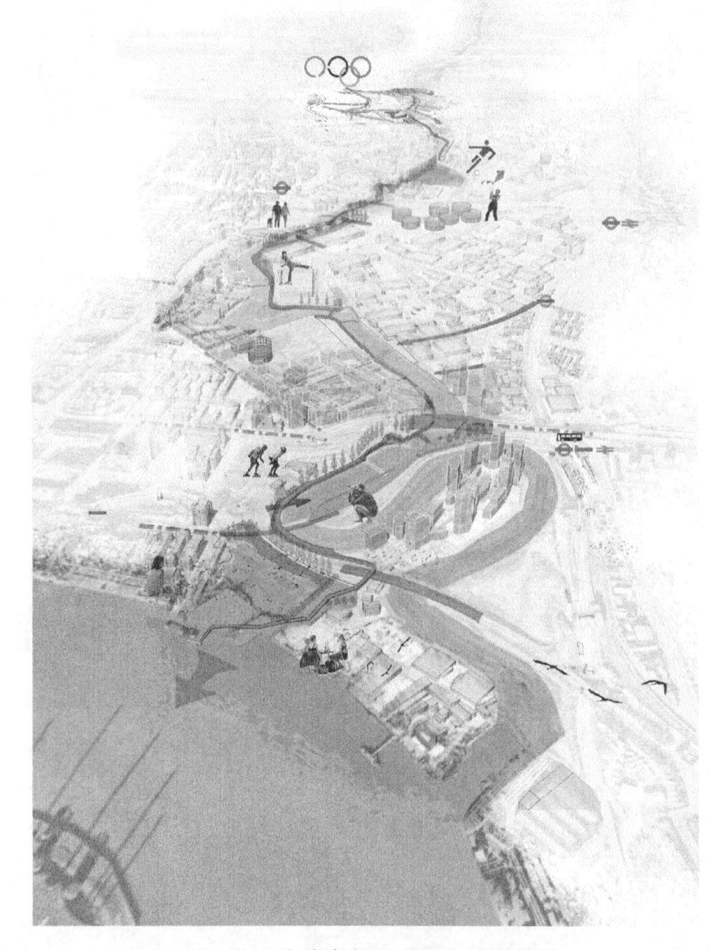

图 26　位于伦敦东部下溧河谷的奥运村*

■ 克服郊区蔓延和景观破碎化的措施：大型城市区域内部的很多问题集中在那些杂乱无序但却在持续扩张的边缘地带，这些地区内部分布着各种孤立的碎片化居民区，它们相互之间差异巨大，在功能和社会关系方面与外部隔离。对于这些因城市蔓延形成的景观破碎化地区，城市设计需要承担三个任务：控制景观破碎化的程度，开发具有一定独立性的郊区小镇以防止出现完全依赖外部的"居住小区"，提升现有郊区地带景观的品质。

■ 发展绿色城市：2003 年，欧盟规划理事会发布了《新雅典宪章》(*New Athens Charta*)。作为引导欧洲城市可持续发展的重要文件，其明确提出转变"现代主义城市设计"的方向。人们在发展可持续城市方面提出了一系列新的构想，包括发展更高的密度、减少景观破碎化、发展城市绿地空间、兴建更多的太阳能板和保温建筑、追求更多的功能混合和城市多样性。

■ 应对气候挑战和发展可持续的交通模式：未来的大型城市区域需要发展新型的交通空

间和交通工具。可持续的机动性将变得日益重要,人们需要发展节约用地、资源友好且更具有社会包容性的出行方式。气候变化和能源危机改变了人的行为方式:除了汽车之外,其他交通工具也得到越来越多的使用(图 27)。

图 27　柏林自行车道的建设(Léon Jaussely und Charles Nicod)*

■ 对战略性规划的推崇:具有综合性特点的“大规划”重新获得了广泛吸引力。战略规划工具以城市和区域当中的土地利用与区位条件为对象,目的是对地区内部各种与人类活动有关的空间关系进行综合调控,使地块开发过程能够保持长期稳定。鉴于其战略性意义,在进行空间结构方面决策的时候,战略规划工具必须同时考虑社会、经济和生态等多方面的要求(图 28)。[5]

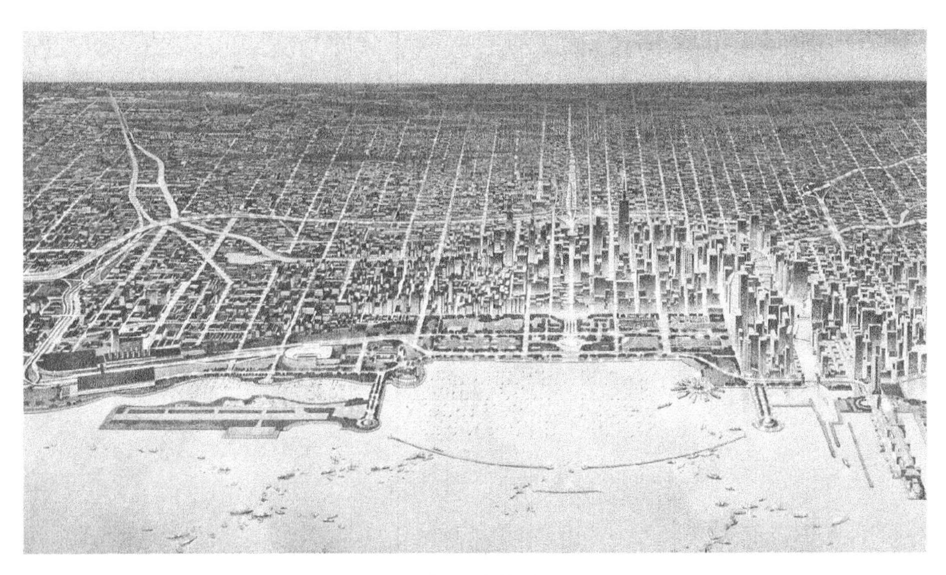

图 28　芝加哥规划 2020*

1.7 结论

通过举办“城市愿景 1910 ｜ 2010”展览，人们对百年前后西方国家的城市设计主题进行了总结。对于今天的城市来说，面临着气候变化、经济全球化，1910 年的展览仍然使人认识到，一个具有现代化水平、技术进步的城市，必须是有活力并且具有集体认同和目标的城市，100 年前的乐观态度也将鼓舞着当代的人们塑造充满意义和价值的城市。[6]

作为 100 年前的先驱，1910 年的大柏林竞赛暨国际城市设计展成为推动城市设计学科发展的一次重要事件，同时进一步促进了城市设计学科的国际化。1910 年的展览本身也是一场庆典，宣告着在现代社会发展中，城市崛起成为重要的力量。在《建筑世界》(Bauwelt)杂志 108 期中，维尔纳·海格曼指出，大都会并不仅仅是现代交通和运输工具的产物，而是“现代社会化的表现，依靠劳动分工，提供了前所未有的智力与体力合作的可能性。城市就是行动者！”[1]

说明：

* 以上图片来自参考文献[1] Bodenschatz, H (ed.). Stadtvisionen 1910/2010.

参考文献

[1] Bodenschatz H. Stadtvisionen 1910/2010：100 Jahre Allgemeine Städtebau-Ausstellung in Berlin[M]. Berlin：DOM publisher, 2010.

[2] 易鑫，吴莹.百年前/后的城市设计学科发展　从柏林“城市愿景 1910/2010”展览谈起[J].建筑与文化，2013，12(11)：138-143.

[3] Frick D. Theorie des Städtebaus[M]. Thübingen：Ernst Wasmuth Verlag, 2010.

[4] 易鑫.“尊重现状的城市发展”：当代柏林的城市更新实践[J].建筑与文化，2014(5)：146-151.

[5] Lynch K. Good city form[M]. Cambridge, MA.：MIT Press, 1985.

[6] 易鑫.“城市愿景 1910|2010：柏林·巴黎·伦敦·芝加哥·南京”城市设计展[J].建筑与文化，2016(12)：1-19.

第三部分

城市设计与更新实践

重塑人的关系是历史地段空间再生的关键因素
——南京老城小西湖历史风貌区保护与再生实践

Reshaping human relationship as the keystone of historic area regeneration: The conservation and regeneration practice of the Xiaoxihu Historic District, Nanjing

韩冬青　董亦楠
Han Dongqing　Dong Yinan

摘　要:作为一种复杂的空间实践和社会实践活动,历史地段的保护与再生不仅是物质空间问题,同时涉及不同角色的立场和权益。重塑人的关系是历史地段空间再生的关键因素,而产权则是其中的重要切入点。在南京老城小西湖历史地段保护与再生工作中,我们通过产权信息的梳理与呈现、配合征收进程的动态规划、配合规划管控的更新图则以及结合居民意愿的设计策略,重塑了相关参与者之间的关系架构,探索了历史地段空间再生的新路径。
关键词:历史地段;保护;再生;角色关系;产权

Abstract: As a complex spatial and social practice, the conservation and regeneration of historic areas involve not only the design of physical space, but also the positions and interests of different participants. Reshaping human relationship is the keystone of historic area regeneration, and property rights are an important entry point. In the conservation and regeneration project of Nanjing Xiaoxihu historic district, through the presentation of property information, the dynamic planning cooperated with expropriation, the renewal plans to assist planning management and control, and the design strategy combined with residents' willingness, we reshape the relationship between the related participants, and explore the new path of historic area regeneration.
Key words: historic area; conservation; regeneration; participants' relationship; property rights

1　背景

南京是中国"四大古都"之一,素有"六朝古都""十朝都会"之称。老城南地区更是具有历史文化代表性意义的区域，其中包括4个历史文化街区、7个历史风貌区，承载了极其丰富的历史记忆，是南京城市发展历程的一个缩影。过去的几十年间，南京出台了一系列保护规划和实施政策，探索老城改造策略，成就斐然，也有争议。老城历史地段的保护与发展既是空间实践，更是一种复杂的社会实践。不同的角色呈现不同的认知和诉求，并伴随一系列的冲突和矛盾。“保护 vs.发展”这个老生常谈的问题始终是社会各界关注和争论的焦点，也引发政府部门和学界的不断反思。

小西湖地段位于南京老城南门东地区，紧邻内秦淮河东段，北近夫子庙历史文化街区(图1)。小西湖地段(又称“大油坊巷历史风貌区”)是《南京市历史文化名城保护规划(2010—2020)》确定的22处历史风貌区之一，六朝以来多有文人雅士富商聚居于此，历史人文资源丰富，是南京老城南地区为数不多的、较为完整保留了明清风貌特征的传统住区之一。现代以来，小西湖地段的历史价值逐渐淹没于激增的人口、复杂的产权关系和无序的建设之中，地段内房屋破旧、居住拥挤、公共设施短缺，几处重要文保和历史建筑年久失修，居民普遍缺乏自我更新能力。据2016年的调查数据，不足5公顷的用地内，容纳了近1 200户居民(其中866户持有产权证)、25家工企单位，居住人数超3 000人，人均居住面积仅10余平方米，严重低于《江苏统计年鉴2016》中南京人均住房面积36.5平方米。2015年，小西湖地段以秦淮区重点棚户区改造项目备案立项，由南京历史城区保护建设集团(简称“历保集团”)负责，探索历史地段“城市修补、有机更新”新模式，该项目面临“风貌区保护”与“棚户区改造”双重任务，其实质是兼顾历史保护与活力再生。2015年暑假，南京市规划局组织东南大学、南京大学和南京工业大学共同参与“小西湖规划研究志愿者活动”，探索“小尺度、渐进式”老城保护与更新方法。2016年底，受历保集团委托，东南大学韩冬青教授团队正式开展小西湖修建性详细规划编制以及市政设施和部分示范性院落的设计工作。

图1　南京小西湖地段区位

(来源：百度地图+笔者自绘)

该项工作涉及政府管理、政策制定、城市规划、城市设计、建筑设计、文物保护、社区营造、资本运作等方方面面，是众多不同角色共同参与的复杂性合作共轭系统。在小西湖研究和实践过程中，不同阶段参与者大致可分为五类：其一，政府职能部门。包括市区两级政府和规划、土地、房产、建设管理等职能部门，其代表着城市和民众的公共利益，并承担与其职能相对应的管理服务职责。街道办和居委会等社区基层组织承担政府与居民沟通的桥梁作用,并组织居民开展社区营造活动。其二,国企建设平台和经营者。历保集团作为秦淮区属国企,全面负责小西湖地段征收、历史建筑修缮、公共设施建设和招商运营的整个流程。同时,小西湖地段也因其特定的区位优势而吸引众多经营性企业的关注。其三,居民,又分为私房产权人、公房承租人以及实际租住人等。小西湖地段人口密度高,老龄化、边缘化严重,居民对于去留以及生活环境有着不同的需求。其四,研究与设计团队。包括社区规划师、建筑师、市政工程师、室内设计师等,其中社区规划师全面负责风貌区历史与现状研究、保护与再生规划设计以及与各方参与者的沟通和协调。其五,市民公众和游客。他们关注南京老城南保护与再生进程,并期待体验历史文化的传承和新的城市活力。

不同的参与角色对历史地段保护与再生过程提出不同诉求,其责任、权利互相交织,形成复杂的关系网络。几十年来老城历史地段保护与再生的经验和教训都表明,空间实践是社会实践的一种外在显现,而社会实践必须坚持既见物又见人。如果忽视了物质空间形态背后的人的因素,空间实践就会难以推动,或徒有其表,甚至走向初衷的反面。显然,重塑人的关系是历史地段空间再生的关键因素,而产权则是这种关系重塑的重要切入点。

2 产权作为历史地段保护与再生空间实践的基础

2.1 产权信息:梳理与呈现

产权即财产权利(property rights),指产权主体拥有的与财产有关的权利的集合,是一定社会的人与人之间财产关系的法律表现。在城市形态学研究中,产权经常被作为地块(plot)的一种隐藏属性。美国学者穆东(Ann Vernez Moudon)认为地块即一组产权单元[1];英国学者克罗普夫(Karl Kropf)通过对地块、产权与行为的分析,提出地块作为一种产权和控制性元素,根植于人类行为和相互关系中[2];中国学者郭莉、赵辰用“地界”对应西方城市形态学中的“plot”,以指代城镇中以产权为依据的土地划分单元,并指出地界是中国古代私有制为基础的土地制度的产物[3]。

在中国现行制度下,不动产权分为土地使用权和房屋所有权。1949 年以前的封建社会和民国时期普遍实行土地私有制,国家和个人都可以拥有土地并自由买卖。业主持有的房地产契中注明了土地四至和面积等信息,通常一块土地即一户人家,土地与房屋产权是合一的。1949 年以后,土地制度开始由私有制转向公有制。首先进行改造的是农村土地。1951 年始,各级政府依法对城市居民的私有土地和房屋进行登记,私房户主凭民国时期的房地产契换取《房地产所有证》,所有证以宗地为单元,内含地籍图,代表此土地为私有财产。随着城市人口增长与住房短缺的矛盾日益突出,1956 年,中央政府发布《关于目前城市私有房产基本情况及进行社会主义改造的意见》,私房改造运动针对城市私有出租房屋设置改造起点(例如上海、南京为 150 平方米,北京为 225 平方米),超出的面积由国家统一“经租”,即在一定时期内给房主以固定租金,将经租房产权逐渐由私有转变为国有[4]。该意见还要求,“一切私人占有的城市空地、街基等地产,经过适当的办法一律收归国有”,但不涉及城市私有房屋基地的征收或国有化。自此，土地产权和房屋产权开始分离。1966年始，许多私房业主受到冲击，迫于形势将私房交公。1969年城市居民下放农村时，下放户的私房全部由政府低价收购。同时，地方政府将所有城镇土地收归国有。“文革”结束后，重新落实私房政策，发还其间无偿交公的房产，补偿低价收购的下放户私房房价。1982年“城市的土地属于国家所有”明确写入宪法，1988年宪法修正案明确“土地的使用权可以依照法律的规定转让”。

南京小西湖地段内的产权演变存在两条线索。首先是土地产权，从1936年的117个私有产权地块，逐渐演变为现状216个国有产权地块，其使用权分别属于私人、单位和南京市秦淮区房产经营公司(国家)。80年间，产权地块数量几乎增加了一倍。将两个时期的地籍图进行叠合，可以发现很多地块边界线基本保持不变，最主要的变化出现在大地块内部的细分以及局部地块边界的收缩(图2、图3)。其次是房屋产权，原本与地权对应的房屋产权从私有产权演变为私房、系统公房(单位)和直管公房(国有)并存的混杂状态(图4)。从土地使用权与房屋所有权的对应关系看，大多数国有和单位持有的产权地块对应公房，原本的一个地块甚至一栋建筑被分租给几户甚至十几户承租人；私人持有地块对应于私房，但是随着土地使用权的继承、买卖等不断细分，同一建筑物又可能被划分到多个产权地块内；此外还有少量存在历史遗留问题的地块，在同一产权地块内公房与私房共存。

图2　小西湖地段1936年产权地块分布图
(来源：根据1936年南京市地籍图绘制)

图3　小西湖地段2018年产权地块分布图
(来源：根据2018年用地红线权属范围图绘制)

从房屋所有者与使用者的责权利关系看，公房产权属于国家或单位，承租人没有修缮的权利和义务。但是作为一种社会福利的延续，公房一直实行限制性的低租金制度，房租并不足以支付标准的房屋维修费用。因此作为产权人的地方房产经营公司和各单位只能按低标准保证房屋的基本安全。对于私房住户，由于大部分私房面积很小，与其他毗邻住户共享同一栋建筑或院落，难以清晰界定实际使用范围，这也妨碍了产权人对私有房产进行有效维护和修缮。可以说，复杂的产权关系是中国城市历史地段物质衰败的重要原因之一。对于土地和房屋产权信息的梳理与呈现是保护与再生规划设计与实施建设难以回避的前提性工作基础。

图 4　小西湖地段房屋产权分布图

（来源：小西湖修详课题组①）

2.2　动态规划：配合征收进程

历史地段人口密度大，人均居住面积远低于城市平均水平，在风貌保护及其建设强度的指标约束下，适当的人口疏解无疑是必要的。2017 年 3 月，历保集团下属的征收公司和夫子庙街道共同组建小西湖地段征收团队，5 月份正式展开民意调研和征收工作。为尽可能平衡居民意愿与改造可实施性，小西湖项目采用"自愿、渐进"的模式，以"院落或幢"为单元进行搬迁和修缮，在充分尊重民意的情况下，分步对全部居民同意交房的"院落或幢"进行改造。这种征收模式与以往将整个风貌区划为征收范围、全部搬迁的征收模式相比，主要的特点在于"自主搬迁"和"动态搬迁"：前者指征收过程充分尊重居民意愿，以自愿搬迁作为征收的基本原则；后者则是由于调研中发现居民的参与程度和意愿处于不断的变化中，通过与居民的持续沟通确定去留，逐步实施征收计划。

为配合这种"自愿、渐进"的征收模式，研究与设计团队在类型学地图（typological map）②上叠加产权信息，以红色实线绘制土地产权边界，并且标注每一户居民的产权人姓名，体现房屋产权分布。叠加产权信息的类型学地图清晰展现出每一个院落、每一栋建筑以及其居民信

① 小西湖修详课题组为简称，全称为"小西湖修建性详细规划课题组"。

② 类型学地图是一座城市、街区或者地块内所有建筑物的地面层平面图，曾是意大利城市形态类型学研究中的重要工具。

息。继而，把类型学地图拆分成12个部分，规划设计团队成员负责与各征收小组具体对接，逐一核对产权信息与实际居住状况，工作人员即时标注自愿搬迁的居民，动态记录征收工作的全过程。2017年4月起，设计团队与征收团队共同工作，将小西湖地段全体居民的意愿和具体搬迁情况落实于图纸，绘制出不同阶段的居民搬迁意愿分布图。这些基于类型学地图绘制的新图纸，不仅是规划设计团队、征收团队与居民之间沟通联系的桥梁，也成为政府部门确定最终征收范围和征收方式的有力支撑(图5)。

图5　居民搬迁意愿

(来源：小西湖修详课题组)

在兼顾历史保护、公共利益和居民意愿的前提下，经规划部门、历保集团、设计团队和居民的一致认同，明确已完成搬迁的12个地块作为一期实验性实施项目，另有6个即将完成搬迁的地块作为二期项目。这些项目以自上而下的方式由国企建设平台负责实施，为整个地段提供公共场所和社区服务功能，提升环境品质和生活水平，同时也将促进留守居民自下而上的更新进程(图6)。

2.3　更新图则：配合规划管控

在规划实施阶段，小西湖地段“小规模、渐进式”的空间实践同样遭遇了一系列政策瓶颈。对于国土管理部门，“自愿、渐进”征收的产权地块面积普遍较小，且布局分散，难以按照传统的工作模式对整个地段进行土地收储、出让或划拨，急需探索小尺度地块“渐进式”操作方法。对于规划管理部门，关键则在于如何制定与此相适应的规划管控办法和审批流程，确保多元主体

图 6　实验性实施项目

（来源：小西湖修详课题组）

在渐进式的改造过程中，遵循共同的规划意图和建设规则。

规划设计团队通过与土地及规划管理部门反复研讨和沟通，最终提出与产权信息和征收进展相适应的小尺度分层级管控体系：在小西湖地段整体规划原则的指引下，再分出“规划管控单元”和“微更新实施单元”两个层级。其中，规划管控单元共 15 个，是指由街巷围合的地块系列，明确用地红线，规定边界特征、退让要求及出入口范围。微更新实施单元共 127 个，是设置详细规划指标、制定详细管控要素的基本单元，其划分基于现状 216 个产权地块，包括四种情况：延续现状产权地块；整合居民自愿搬迁、相邻且有条件进行联合改造建设的系列产权地块；基底面积难以承载南京市平均居住面积且肌理极度零碎的地块合入相邻地块；部分相邻地块被同一栋建筑占据，无法独立改造而合并（图 7）。继而，对 15 个规划管控单元分别编制微更新单元图则，为其中各个实施单元的改造建设行为提供具体的控制和引导要求。这些管控要素包括两方面内容。第一是对各实施单元的边界、规划指标、用地性质进行界定，并对业态和既有建筑的处置方式提出引导。第二是边界类型管控，根据老城南建筑类型的诠释性研究成果，对院落式建筑布局的山墙（院墙）和檐墙界面分别提出不同的管控要求。历史地段内保留和新建建筑山墙的位置和方向应与历史肌理一致，并保持连续度；而檐墙界面则允许其在地块红线内适当退让，同时规定连续沿街立面长度不宜超过 15 米，以维护传统街巷风貌特征；鼓励街道活力界面，采用“前店后住、下店上住”等模式，引入社区公共服务功能。依循这一规则体系，无论未来的改造主体是产权个人、开发商还是政府，均可根据图则要求对相应地块或地块系列进行自下而上的改造活动。同时，管控图则也是各相关部门和角色对设计、建设和运维过程进行论证、监督和建成后评估的基本依据（图 8）。

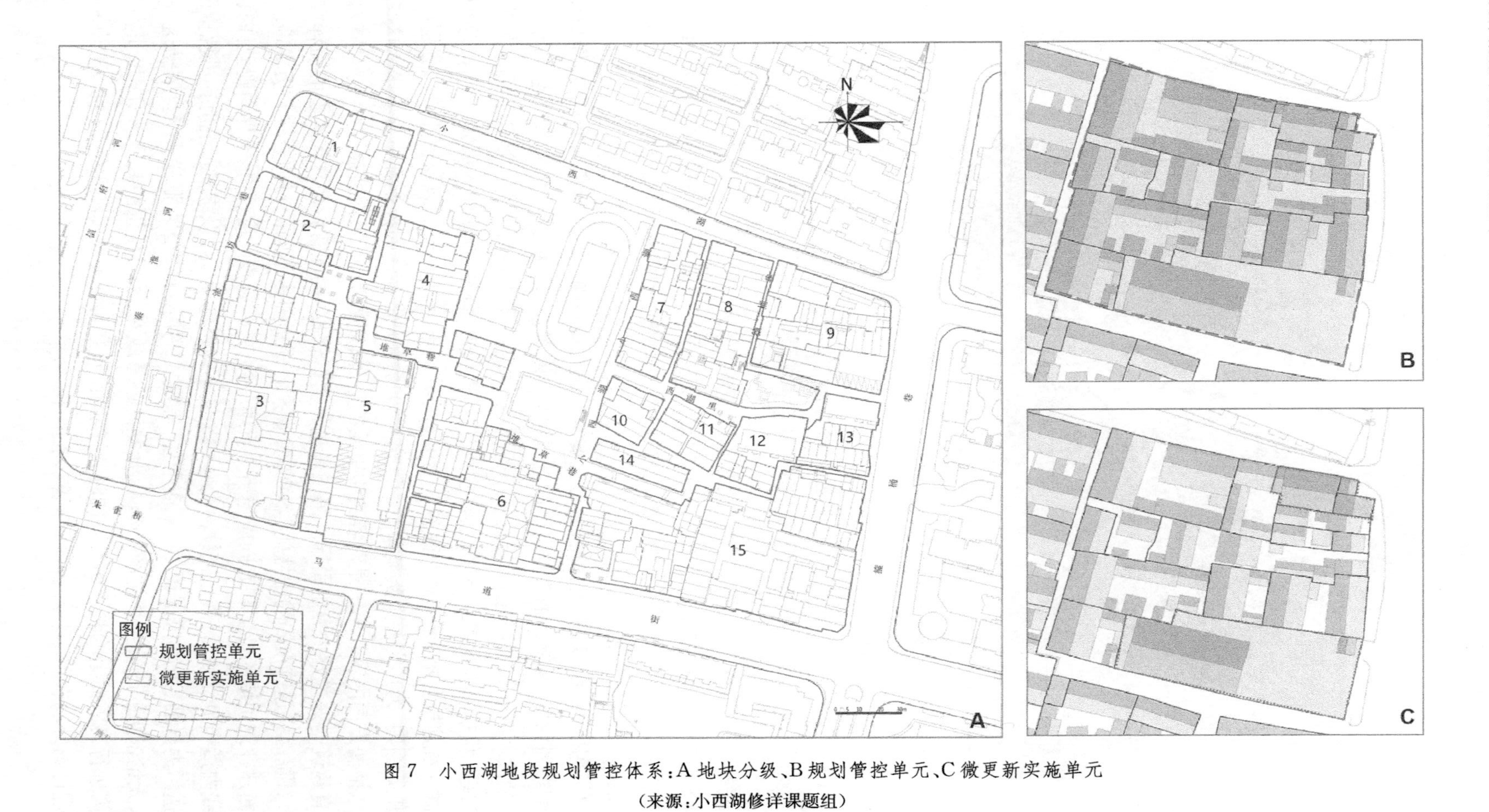

图 7　小西湖地段规划管控体系：A 地块分级、B 规划管控单元、C 微更新实施单元

（来源：小西湖修详课题组）

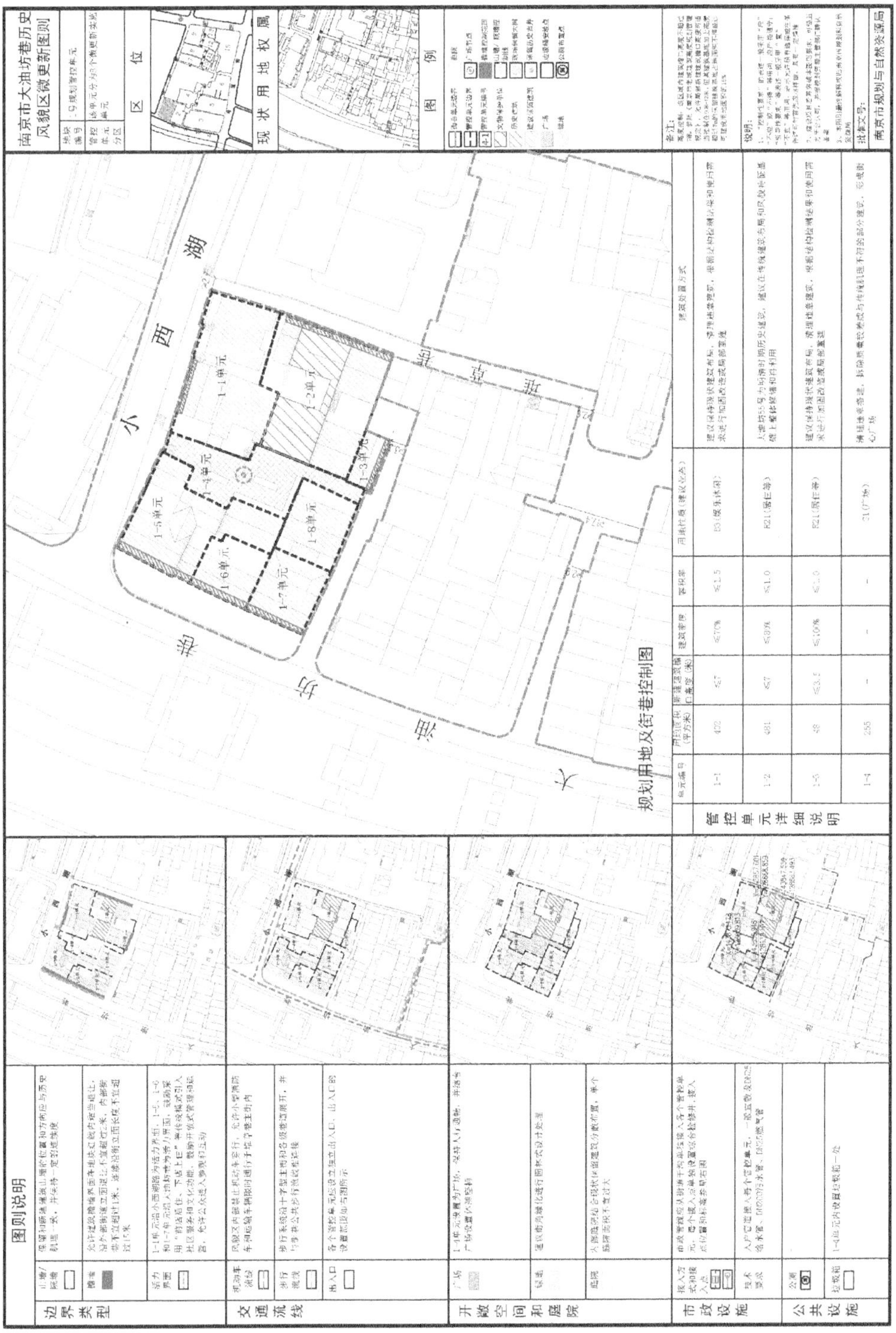

南京市大油坊巷历史风貌区微更新图则

地块编号	1号规划管控单元
管控单元分区	该单元分为8个微更新实施单元

区位

现状用地权属

图例

图则说明

类别	要素	说明
边界类型	山墙/院墙	保留和新建建筑山墙的位置和方向应与历史肌理一致，并保持一定的连续度
	檐墙	允许建筑檐墙界面在地块红线内适当退让，沿外部街道立面退让不宜超过2米，内部街巷不宜超过1米。连续沿街立面长度不宜超过15米
	活力界面	1-1单元沿小西湖路为活力界面，1-5、1-6和1-7单元沿大油坊巷为活力界面，鼓励采用“前店后住、下店上住”等传统模式引入社区服务和文化功能，鼓励开放式管理和运营，允许公众进入参观和互动
交通流线	机动车流线	风貌区内部禁止机动车穿行，允许小型消防车和运输车辆限时通行于堆草巷主街内
	步行流线	步行系统沿十字型主街和各级巷道展开，并与步行公共步行流线相连接
	出入口	各个管控单元应设立独立出入口，出入口的设置范围如右图所示
开敞空间和庭院	广场	1-4单元设置为广场，保持人行通畅，并结合广场设置休憩座椅
	绿地	建议街角绿化进行园林式设计处理
	庭院	大部庭院结合现状保留建筑分散布置，单个庭院面积不宜过大
市政设施	接入方式和接入点	市政管线应从街道干沟单独接入各个管控单元，每个接入点单独设置综合检修井，接入点位置和标高参见右图
	技术要求	入户管道接入每个管控单元，一般应敷设DN25给水管、DN200污水管、DN25燃气管
公共设施	公厕	-
	垃圾箱	1-4单元内设置垃圾箱一处

规划用地及街巷控制图

管控单元详细说明	单元编号	用地面积(平方米)	新建建筑檐口高度(米)	建筑密度	容积率	用地性质(建议业态)	建筑处置方式
	1-1	402	≤7	≤70%	≤1.5	B3(娱乐休闲)	建议保持现状建筑布局，清理违章建筑，根据结构检测结果和使用需求进行加固改造或局部重建
	1-2	481	≤7	≤80%	≤1.0	R21(居住等)	大油坊55号为明清时期历史建筑，建议在传统建筑布局和风貌特征基础上整体修缮和再利用
	1-3	48	≤3.5	≤100%	≤1.0	R21(居住等)	建议保持现状建筑布局，清理违章建筑，根据结构检测结果和使用需求进行加固改造或局部重建
	1-4	255	-	-	-	G1(广场)	清理违章搭建，拆除质量较差或与传统肌理不符的部分建筑，形成街心广场

批准文号：

南京市规划与自然资源局

图8　小西湖地段1号规划管控单元微更新图则

(来源:小西湖修详课题组)

2.4 设计策略:结合居民意愿

自2015年始,小西湖地段的保护和再生实践始终关注居民意愿。设计团队成员和历保公司的人员带着规划设计方案走进社区,与居民交流设计构想,倾听民声,并建立了线上微信交流平台“社区建筑学”(图9)。在一期实施项目的设计中,结合居民意愿和搬迁情况,设计团队灵活采取不同的策略。例如,快园地块在历史考据的基础上,综合用地的交通组织、历史要素和功能安排,形成有历史意味的、供社区居民活动的公共空间和标志性社区景观(图10);平移安置住房地块将安置无意外迁且现有居住条件较为恶劣的原住民,设计结合街区的传统风貌对既有建筑进行优化提升改造,最大限度地提高平移安置住户的生活质量和土地使用价值(图11);共生院地块对一个保留了两户原住居民的特色院落进行适应性设计,打造原有居民与新生业态共生共存的示范院落(图12)。诸如此类的参与性设计和建设仍在持续。

图9 走进社区
(来源:小西湖修详课题组)

图10 快园公共空间
(来源:小西湖修详课题组)

图11 平移安置住房
(来源:小西湖修详课题组)

图12 共生院
(来源:小西湖修详课题组)

3 空间实践中角色关系的重塑

在小西湖历史地段的空间再生实践中,我们时常陷入一种多主体相互碰撞冲突、莫衷一是的尴尬境地。每个角色呈现不同的姿态,表达不同的意愿,所有人都怀揣对未来的美好期待,但具体的期待却不尽相同。不同的角色立场都有其具体的责任、权限和利益,问题在于如何能汇集方向一致或相近的力量,又避免消极的不作为甚至伤害。我们显然需要一种能沟通各方角色，兼顾并平衡各方诉求的交互平台，并借助这样的平台，重塑空间实践乃至社会实践中诸多角色之间的良性共轭关系。

"五方平台"是在小西湖实践过程中提出的，针对历史地段保护再生规划以及各独立设计建设项目的联合磋商、管理和服务的平台。平台主体包括政府职能部门、街道办和居委会、国企建设主体和相关市场主体、社区居民以及社区规划师。其中，政府职能部门包括区政府、市规划和自然资源、房产、建设、文旅等部门，负责五方平台的建立、管理和建设方案的批准。街道办和居委会负责居民的宣传和联络，以及社区营造的开展。社区居民指提出更新申请或与更新项目相关的临近居民。保护和建设单位明确改造方案的可实施性以及新增业态的合理性。社区规划师由市规划资源部门和国企建设平台联合委任，监审规划意图的落实和管理，参与更新项目策划、设计、建造以及后续发展维护的全过程，以专业视角参与协调不同权益方。五方主要参与者共同商讨并审批规划和建筑设计方案，并在后期施工和使用过程中起到监督和协调作用。基于五方平台的微更新工作流程如表1所示：

表1 微更新工作流程

1) 按照产权类型疏解人口	按照产权类型分类开展，原则上以公房腾退、私房自愿腾迁、厂企搬迁等方式进行；建立五方平台(政府职能部门、街道办和居委会、社区居民、国企建设主体和相关市场主体以及社区规划师)和社区规划师制度
2) 提交更新申请	以产权人或承租人为基本单位，依据更新单元图则向社区规划师提出更新申请，并由社区规划师签字、公示
3) 提供更新条件	市规划与自然资源局根据批准图则要求，向申请人提供规划指标、业态功能、市政管网结构等更新技术条件
4) 更新方案设计	申请人依据更新技术条件，委托社区规划师或者其他在国企建设平台备案的设计单位，编制更新设计方案和管线实施方案，并征求相邻产权人意见
5) 五方平台审核	由社区规划师组织召开五方平台会议，形成书面意见(含消防及节能审查)。若较更新前无产权面积变化和土地性质改变，经规划资源部门备案后，进入施工申请流程。若涉及，则须完善土地登记手续
6) 施工建设组织	申请人在五方平台自选或抽选的施工单位资源库中，选择施工单位，并报区住建局批准后组织施工。施工过程中应减少对周边居民的影响，社区规划师定期现场巡查，给予技术支持
7) 竣工验收	施工完成后，由社区规划师组织五方平台联合进行竣工验收

(来源：小西湖修详课题组)

随着规划设计和建设工作的展开，通过对不同地块和房屋的产权类型、居民意愿的梳理以及使用功能的改变，结合不同更新类型，重新塑造了产权主体、更新主体、使用主体之间的关系。例如，自愿腾退的公房和自愿腾迁的私房被国企建设平台统一收购，或改造为公共服务建筑，提升社区居民生活品质并为游客提供旅游信息服务，或与市场经营者联合更新，为街区提供商业服务。再如，私房产权人可原址居住或自由出售给新的私房产权人，继而选择自我更新并继续居住，或将私房出租给国企建设平台或经营者，引入商业或文化功能，服务当地居民或游客。在此过程中，社区规划师既是专业设计的贡献者，也是参与协调的中介和服务者。

4 结语

既有的经验和教训已经反复证明：无论是推土机式的大拆大建，还是某种单一力量的一厢情愿，都难以回应老城历史地段保护与再生所面临的复杂问题。所谓“小尺度、渐进式”模式，不仅仅是一种物质空间操作的理念和策略，其背后更是不同角色之间碰撞、交流、理解、互动、合作的机制与进程。小西湖实践是多重角色共同参与的大合唱，也将是不断改写剧情且难以谢幕的持续进程。外在的物象与内在的情理、可视的场所环境与不可视的政策机制，透过人与物的彼此纠缠和相互成就，谱就在矛盾与互动中不断探索且生生不息的人居进行曲[5]。

致谢：

本文在研究、实践和写作过程中，得到东南大学建筑学院小西湖修详课题组鲍莉、邓浩、沈旸、李新建，南京市规资局吕晓宁、李建波，历保集团范宁、黄洁、张剑、叶晓明等的帮助和指导，在此表示感谢。

参考文献

[1] Moudon A V. Built for change[M]. Cambridge, MA: MIT Press, 1986.

[2] Kropf K. Plots, property and behaviour[J]. Urban Morphology, 2018, 22(1): 5-14.

[3] 郭莉，赵辰.地界在中国传统城市肌理研究中的意义[C]//中国城市规划学会城市规划历史与理论学术委员会.2014年第6届城市规划历史与理论高级学术研讨会暨中国城市规划学会城市规划历史与理论学术委员会年会论文集.泉州：中国城市规划学会历史与理论学术委员会，2014：2-13.

[4] 南京市地方志编纂委员会.南京房地产志[M].南京：南京出版社，1996.

[5] 董亦楠，韩冬青，沈旸，等.适于传统街区保护再生的“类型学地图”绘制与应用：以南京小西湖为例[J].建筑学报，2019(02)：81-87.

年轻的城市设计师：让德国的儿童和青年参与城市发展

Young urban designers: Involving children and youth in urban development in Germany

［德国］安娜·尤莉安娜·海因里希
Anna Juliane Heinrich

城市规划应该是以人民为本、市民参与的沟通过程。自1970年代以来，德国和世界其他许多国家已经把这种态度视为理所当然。最初，交流性规划的讨论和实践侧重于“参与”的形式，即市民在决策中拥有发言权选项，不过这些选项是由政治和行政部门开放和协调的。此后，强调公共和私人领域之间属于两极对立关系的基本假设已经消散，人们对城市发展有了更为多元化的理解。现在，城市发展被理解为政治、行政、经济、市民社会和中介利益相关者的不同逻辑与利益之间的相互作用过程。

对于城市规划及其利益相关者的复合性问题来说，笔者将进一步提出以下观点：以人民为本的城市发展不应局限在以城市规划师为重点的方法，尽管他们是考虑“人民”和为人民设计的专业人士。尽管当然应该将“人民”放在规划工作的核心，但是这种方法还不够。以人民为本的设计应采用使“人民”积极参与城市发展的方法。更重要的是，笔者认为城市发展本身就是由市民一起塑造的，规划专业人员和决策者除了应该承认这一点以外，还应该赞赏和促进相关的城市发展。毫无疑问，对城市发展的这种广泛理解给城市规划人员带来了巨大的挑战。这就需要人们进行反思，这将改变我们对城市规划工作、规划专业人士的角色和职责的理解。

在本文中，笔者关注的是年轻人群及其在城市发展中的作用。就德国而言，几十年来一直将儿童作为规划的目标群体。特别是在开发家庭友好型住宅区的背景下，规划专业人员会考虑儿童这一人群。仅在过去的15年中，青少年已越来越多地参与规划过程。而且人们普遍认为青少年从一开始就是规划制定过程的积极参与者。近年来，他们被越来越多地当作是能够独立起到塑造城市发展作用的利益相关者。

在下文中，笔者将区分年轻人群参与城市发展的三种不同模式：国家主导的参与、青年倡导者主导的参与和青年人主导的项目。这将拓宽我们对年轻人参与的理解，并为在不同背景下的市民参与和共同生产开辟一系列机会。再下面将专注于青年倡导者主导的参与，在这种城市发展方法中，可能涉及国家机构，但青年倡导者作为中介利益相关者负有责任；鼓励年轻人并向其赋权，使其担任城市设计师。作为这样一个青年倡导者的例子，笔者将介绍德国的非营利组织“青年建筑城”(YOUTH ARCHITECTURE CITY)。这样做的背景是：一方面，本人就是该组织的积极成员；另一方面，我与同事们一起对“青年建筑城”组织的参与过程进行了研究。为了启发参与过程，我将描述非营利组织如何设计四个阶段的参与过程：探索、设想、设计和交流。接下来还将介绍一个有关年轻人参与游乐场重建的实际案例，并反思年轻人参与的一些成功因素和障碍。最后，将简要介绍由于年轻人参与方式不同，其对城市规划专业人士所产生的影响。

不同的参与方式：国家主导、青年倡导者主导和青年主导

1992年，罗杰·哈特(Roger Hart)针对年轻人的参与树立了一个里程碑：结合雪莉·阿恩斯坦(Sherry Arnstein, 1969)提出的“阶梯式参与”的概念，引入了专门针对年轻人参与的内容，把否定的参与模式分为操纵、“象征主义作秀”等不同情况(Hart, 1992: 8-14)。并且，他界定出各种不同的参与层次，包括年轻人被“指派任务但得到告知”“成年人发起但与孩子共同决策”“儿童发起但与成年人共同决策”等(Hart, 1992: 8)。这些不同的参与层次尽管体现了青年代理机构的差异，但是这些机构各自均具有同样重要的意义，需要结合各自的背景发挥作用(Hart, 2008: 23)。

尽管哈特对参与的不同程度及对不同程度的代理机构的区分让人大开眼界，但长期以来，人们一直维持着对参与的单一认识，而这一直在主导参与的实践。年轻人对城市发展进程的参与仍然主要被认为是由国家发起和执行的活动。在程序和方法方面，仍然基本是采用民主意见形成的经典形式(Jupp, 2008; Percy-Smith, 2010; Heinrich & Million, 2016)。

然而，在过去的15年中，已经出现新的讨论帮助人们对参与形式的更广泛理解(包括儿童参与规划的系统文献综述)(Ataol et al, 2019)。朱普批评说：“参与政策框架通常被理解为涉及相当具体的活动和场所，例如参加在市政厅举办的会议。”(Jupp, 2008: 332)她反驳说：“小规模社区团体的工作可以采用与社区政府直接互动的传统论坛方式，为当地人民的参与和赋权提供强大的基础。”(Jupp, 2008: 332)因此，朱普指出：“可能有必要重新考虑哪些是构成‘参与’的要素。”(Jupp, 2008: 333)与之相对应的是区分不同的“参与模式”(Jupp, 2008: 334)。除了“国家主导的倡议”(Jupp, 2008: 334)之外，“社区主导”(Jupp, 2008: 341)活动也应理解为参与。

珀西-史密斯(Percy-Smith, 2010)进一步推进了这一观点，并强调了年轻人参与的重要性。他指出：“我们需要降低对正式、制度化公共决策过程参与的关注，而应更多地关注人们行动方式的多样性，在日常生活中的行为、贡献和实现自身代理意识的多种方式。”(Percy-Smith, 2010: 119f)具体而言，年轻参与者应“有权在参与性倡议中对议程和程序行使权利”(Percy-Smith, 2010: 110)。他提倡拓展参与的内涵，应当纳入各种以主动的市民身份实施的非正式日常实践(Percy-Smith, 2010: 109f)。

基于这种论述，海因里希和米利恩(Heinrich & Million, 2016)区分了青年参与的三种模式：国家主导的参与、青年倡导者主导的参与和青年主导的项目。国家领导的青年参与被理解为由国家机构发起和管理，年轻人通过该过程参与规划过程。主题、空间设置、程序和方法主要由国家机构来确定。这种自上而下的参与具有以下优点：行政管理部门与政治部门之间的

直接联系确保决策者能够听到年轻人的声音，这可以为各种想法的实施提供丰硕的基础内容。此外，由国家领导的参与方式所涉及的主题和规模是其他两种模式往往无法达到的，例如：规划涵盖了整个城市乃至区域的尺度，规划任务和文件具有高度抽象和长期性的特点。但是，这种模式也存在一个缺点：这种话题可能完全脱离了年轻人的日常生活，无法激起他们的兴趣。反复出现的问题是，年轻人对这种参与缺乏积极性和动力，国家机构与年轻人之间的沟通也存在困难(Heinrich & Million, 2016: 61f and 68)。

青年倡导者"是独立或代表国家的委托机构，促进青年参与的利益相关者"(Heinrich & Million, 2016: 57)。例如，青年倡导者可以是非营利组织或私人的城市规划事务所。青年倡导者的特点在于，尽管有国家机构的授权，但青年倡导者显然对它们所涉及的年轻人有所偏袒。由于它们通常具有引导年轻人参与的经验，因此"与青少年保持亲密关系，熟悉他们的兴趣、需求、沟通形式、实践方式和文化"(Heinrich & Million, 2016: 64)，这有助于形成有意义的参与。尽管这种模式的特征具有自上而下的特点，但事实证明青年倡导者对目标群体非常敏感，在解决问题和激发年轻人方面更为成功(Heinrich & Million, 2016: 64 and 68)。

年轻人参与城市发展的第三种模式是青年主导的项目。这些包括"年轻人各种形式的活动和承诺，以某种方式与城市环境及其市民具有相关性"(Heinrich & Million, 2016: 68f)。由青年领导的项目是自下而上的举措，其特点是年轻人独立行动并负责其活动。年轻人参与他们感兴趣和关心的问题。在它们的倡议中，它们主要集中在微观尺度上，例如重复使用或重新设计单个地块、建筑物或广场。它们倾向于具有"动手能力"特征的中短期活动。对于年轻人来说，这些都是非常积极的、极大的激励，他们的行为坚定不移。我们的研究表明："青年倡议可以为邻里发展，以及正在进行的规划流程做出宝贵的贡献，并可以激发和提升国家主导青年参与的质量。"(Heinrich & Million, 2016: 63)通常，这些项目没有与城市行政部门正在实施的规划过程联系在一起。但是国家机构关注此类项目，并为它们提供支持或者让它们参与到正在进行的更新改造过程都是值得的。然而，这些倡议有时会被人们视为不受欢迎的干扰(Heinrich & Million, 2016: 63-69)。

可以理解的是，此处描述的三种青年参与模式不应被理解为严格区分开的三种计划，而应该被当作不同的领域，且这些领域相互重叠、扩散和融合(Heinrich & Million, 2016: 69)。通过分析不同的实践，我们可以识别出符合两种甚至三种参与模式特征的混合形式。

"青年建筑城"：作为青年倡导者的非营利组织

非营利组织"青年建筑城"(德语：Jugend Architektur Stadt e. V.; http://www.jugend-architektur-stadt.de/english)就是这样一个德国的青年倡导者。该组织的宗旨是促进年轻人参与规划和建设过程，并促进对儿童和年轻人的建筑环境教育。"青年建筑城"通过其活动邀请年轻人参与他们的环境。鼓励并帮助儿童和青少年用他们所有的感官感知建筑、设计、邻里、公共空间、景观、城市和区域，从而(重新)发现和塑造这些空间。自2005年成立以来，"青年建筑城"的成员已经测试并开发了一系列促进年轻人参与和建筑环境教育的方法(Edelhoff et al, 2019)。

年轻人参与过程的概念化：探索、构想、设计、沟通

毋庸置疑，每个年轻人参与城市发展的过程都是独一无二的。每个过程的特点包括：

—— 追求的目标(例如关于不同的行动领域、利益相关者群体)；
—— 主题(例如交通、聚会场所)；
—— 空间设置(例如广场、邻里)；
—— 利益相关者(例如国家官员、商人)；
—— 目标群体(例如年龄段、特定场所的使用者)；
—— 参与模式(例如青年主导、国家主导)；
—— 应用方法(包括各种广泛类型)(Derr et al,2018)；
—— 资源的可获得性(例如工作人员、财务资源)；
—— 时限(例如短期项目、长期过程)。

由于这些方面会影响参与进程,因此所有内容都需要在开始之前或整个参与过程中得到明确解决。考虑到影响参与进程的众多因素,有必要构建参与进程的方向框架作为辅助。因此,“青年建筑城”组织的参与过程通常被概念化为四个阶段:(1)探索;(2)构想;(3)设计;(4)沟通。

探索环境通常是参与过程的第一步。这将鼓励并允许所有参与者了解参与过程所涉及的环境。在年轻参与者熟悉的环境中,这尤其令人兴奋(但同时也充满挑战)。应该选择合适的方法,帮助参与者睁开眼睛,发现日常生活中被忽略的事物。此外,可以选择一些方法来帮助参与者从新的视角来审视环境。除了这些专注于环境感知的方法外,探索的阶段还包括对品质的评估。年轻的参与者应该认可人们为提高生活质量所做出的贡献并发现问题。在此过程中,他们将需要探索自己的需求和对某个场所的需求,以建立评估标准。为了使参与者意识到城市发展中需要在不同利益之间加以平衡,激励年轻人探索其他利益相关者的需求也很有意义。

第二阶段“构想”的基础依赖于探索环境所获得的敏感性。本质上,此阶段是根据参与者的观点来理解、阐述和阐明需要进行哪些更改。基于前一阶段探索的成果,参与者具有一定的潜力,他们可以在此基础上进一步发展自己的想法。这些问题应以他们的愿景提出,并在可能的情况下予以解决。目的是为环境制定替代解决方案和未来。

第三阶段是设计阶段,这其中涉及对一个或多个构想的决策。毋庸置疑,决策必须基于公平和透明的参与过程。一旦做出决定,通常需要将愿景转化为具体措施。为了使愿景可行,还需要进一步的阐述,设计过程的实际效果在很大程度上取决于整个参与过程的性质(请参见上文:追求的目标、主题等)。

最后阶段是沟通阶段,此阶段针对的是外部交流。当然毫无疑问,在整个参与过程中,内部沟通至关重要。前三个阶段的年轻参与者的工作成果将呈现给更多的受众。由此可以追求不同的策略和目标:可以向公众传达年轻人对环境的评估以及他们对未来发展的愿景。这有助于提高社会对年轻人需求的认识。此外,尤其青少年是一群经常充满偏见的群体,展现他们的愿景可以帮助减少各种偏见。但是,沟通阶段的核心问题通常是与决策者打交道,并确保参与过程能够对城市发展产生影响。

很显然,这四个阶段应被作为理想的模型。在参与过程中,各个阶段肯定会重叠或相互合并。但是,在这四个阶段中进行的工作已经证明了自身的价值:模型作为一个框架,在不过度确定的情况下为人们提供结构和指导。

年轻的城市设计师规划和实施柏林游乐场的重新设计

2014 年和 2015 年,“青年建筑城”让年轻人参与规划和实施游乐场的重新设计工作[①]。这项更新项目位于德国柏林一个贫困居民区内部。游乐场对于居住在该地区的年轻人来说至关重要,特别是因为与之毗邻的是两个备受欢迎的机构:儿童休闲中心和青年俱乐部。

在参与过程的第一年,组织者致力于让儿童和青少年参与重新设计的规划。空间的年轻使用者与负责重新设计的景观规划师之间开展对话。第二年,年轻人参与了具体设计以及游乐场座椅家具的建造过程。这两个项目年都是根据“青年建筑城”的四个工作阶段进行构想的,下文将进行相应的讨论。第一年的工作重点更多地放在了探索和构想上,第二年则强调设计和实际实现年轻人想法的阶段。

整个过程始于对游乐场的探索。当地的年轻人了解当前状况及其对该地区未来发展的需求。为了记录探索的结果,参与者拍摄了照片并设计了明信片,明确展现了他们的评估结果。这有助于更多地了解年轻人对游乐场的看法,并产生一种可以传达这一点的产品。第二步,邀请年轻人进一步了解游乐场的所有(潜在)使用者的需求和愿望。为此,他们采访了附近居民,并制作了短片(图 1)。

图 1

在阐述使用者对重建的需求和想法时,探索与构想阶段顺利融合。为了推进这一阶段,“青年建筑城”创建了一个简单的参与式游戏,用于设想和协商游乐场的不同未来。根据年轻人制作的明信片和视频,确定了年轻人在其邻里所忽略的广泛用途。该游戏的目的是阐明这些用途对年轻人的重要性。按年龄组划分,年轻人获得了平均分配的建筑模块,可以使用这些建筑模块给展示出来的各种用途投票。在游戏过程中,这些建筑模块在游乐场上被堆积成塔,以展示参与者的选择倾向。但是,由于空间(和资金)的可获得性限制了游乐场可供实现的项目,因此游戏的症结就集中在,年轻人必须结成联盟,以实现他们最紧迫的愿望。在玩了一个下午的游戏以后,年轻人提出了重新设计的建议(图 2、图 3)。

图 2

---

① 以下内容基于(1)项目文档(JAS e.V. et al, 2019: 74-81),(2)项目网站(http://www.hingucker-jas.de/index.php/berlin)与作为伴随研究的一部分而进行的非营利组织相关成员的访谈和讨论。

从该项目的经验可以明确探索和构想阶段的两个重要要求：首先，青年倡导者应有意识地选择解决各种交流方式的方法。案例中选择了摄影、明信片设计、视频拍摄、口头讨论和营造游戏气氛的方法。通过为参与者提供各种参与方式，促进包容性参与。不同的参与者会对不同的沟通方式感到满意。因此，应尽量考虑广泛的类型，包括语言方式（例如口头介绍、讲故事、采访、写作），视觉方式（例如摄影、拍视频、绘制草图、拼贴）以及触觉和营造氛围（例如建模、游戏）等交流手段。

图 3

其次，青年倡导者在此过程中创造了两个关键时刻，参与者被要求满足其他人的需求：年轻人采访了其他使用者并一起玩了他们彼此协商的游戏。让年轻的参与者发现其他利益相关者的需求是有价值的，从而使年轻的参与者对城市规划如何始终满足不同的需求有一个现实的想法。它使规划过程对年轻人透明，并为达成共识与合作奠定了基础。

设计阶段仅在某种程度上得以实现。在通过参与游戏来确定目标优先级时，确定了设计阶段的各个方面。但是，正如随后将显示的那样，设计是参与过程第二年的关键。

沟通的阶段涉及三个方面内容：第一，年轻人通过视频剪辑制作了一条视频消息，总结了他们向负责重新设计的景观规划师提出的需求。第二，年轻人对新设计的建议被转发给景观规划师。这份结果传递的工作是由代表年轻参与者利益的青年倡导者组织的。第三，年轻人的建议被印在大幅海报上，并展示在游乐场旁边的防火墙上。

沟通阶段的这些方面暗示与一般沟通阶段相关的一个方面：青年倡导者需要确保参与者的声音被听到。在该项目中，“青年建筑城”与景观建筑师进行了协商，以确保参与者的想法对重建过程产生影响。年轻人的想法确实被采纳，并从根本上改变了景观设计师在参与过程之前已经准备好的计划。为了避免出现“象征主义作秀”的结果，青年倡导者必须时刻注意确保参与的结果真正被纳入规划和实施过程。

此外，除了海报展示以外，向更广泛的公众介绍也很值得。这有助于使邻里感受到年轻人的需求。此外，发挥年轻人的积极作用还可以帮助减少人们对青少年的偏见。

游乐场重新开发项目的第二年被专门用于设计和实施。在与负责的景观设计师协商后，大家达成了共识：参与者可以一起设计和建造未来游乐场的一个元素。根据构想阶段采纳的年轻人想法，项目选择了理想的座椅家具。

在前一年工作的基础上，年轻的参与者立即开始了设计阶段。在模型中发展和展示他们的想法，他们详细阐述了这种座椅家具应具有的特征。当然，座椅家具必须在给定的预算范围内具有可行性，并满足建筑主管部门的所有建筑要求。因此，“青年建筑城”负责将年轻人的设计方案转化为可立即实施的设计。

下一步，所有年轻人都以 1∶20 的比例制作座椅家具的模型。这些模型已经包含了将在施工现场使用的所有施工要素。该步骤用于模拟后面的建造过程，以便参与者可以想象实现过程(图 4、图 5)。

图 4

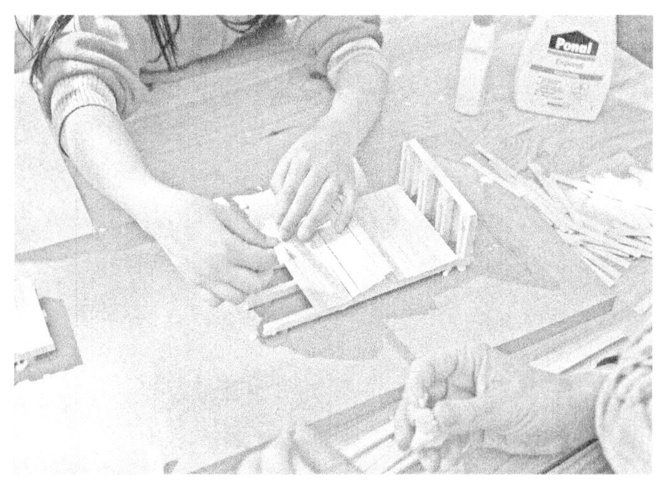

图 5

参与过程的亮点是建立一个参与式建筑工地，用来建造座椅家具。在具有教学经验的建筑专业人士的引导下，儿童和青少年设法在游乐场上搭建木制座椅家具。尽管施工过程和完成的座椅成果已经可以传达出年轻人在邻里中的参与情况，但人们还是另外举行了公众开放仪式，以庆祝年轻人的成功并向更广泛的公众传达他们的宝贵贡献(图 6、图 7)。

图 6

图 7

从该项目示例中可以得出一些经验：在设计阶段，可能有必要调整参与者的建议，以使其切实可行。但是，青年倡导者对此必须非常注意，以便使年轻人继续认同这些是他们的成果。因此，"青年建筑城"邀请年轻人为最终设计制作模型。通过此过程，年轻人可以更深入地了解构造细节，并解释为什么设计是这个样子。更重要的是，这也可以理解为建筑环境教育，它使参与者对设计和建造的过程有一个总体的了解。

参与过程的真正特色之一是参与式建设工作。“青年建筑城”的理想模型通常并不包含实施阶段，毕竟在城市发展过程中，很少有参与活动能够直接发生在建设活动中。不过这个项目仍然是年轻人参与城市发展过程，并取得某种可见成果的成功因素。在“青年建筑城”的其他项目中，采取了截然不同的形式：在参与过程结束时，“青年建筑城”经常激励年轻人留下一些信物，作为对他们曾经经历参与活动的纪念。这有可能是一个海报，并将其挂在参与该过程的机构中；也有可能是年轻人将自己制作的模型带回家，这将支持年轻人牢记他们的成就，证明他们可以有所作为和并改变事物。例如，为了向更广泛公众展示参与的可见性，可以通过临时干预措施吸引公众的注意力并介绍年轻人的需求和想法，或者通过公开展示成果或请当地媒体报道出来。例如，为了向决策者展示可见性，则可以总结年轻人对城市发展主张的宣言，并在公共活动中将这份宣言移交给政客。

结论：让年轻的城市设计师参与的多种方式

在本文中，笔者认为以人民为本的城市规划应包括将年轻人视为伙伴的参与，并采取多种形式。作为专业人士的城市规划师，不仅应该为年轻人规划，还应该与年轻人一起进行规划，并赞赏年轻人推动的城市发展。考虑到这肯定会给城市规划人员带来巨大挑战，因此本文提出了不同的参与方式，这些方式说明了年轻人如何作为城市设计师参与其中。在罗杰·哈特(2008)提出的年轻人参与阶梯上，笔者提出了一个以利益相关者为中心的观点，着眼于由谁负责不同参与模式的问题。国家主导的参与、青年倡导者主导的参与和青年主导项目的模式说明了青年人的知识、需求和思想如何塑造了城市发展。这些模式为规划专业人士提供了不同的机会，使他们了解如何进一步实现、促进、强化或引导城市发展。

本文将重点放在青年倡导者主导的参与和非营利组织“青年建筑城”的工作上，以提供见解和实用的“如何”指导。案例清晰地展示出，年轻参与者可以发挥哪些影响力，以及青年倡导者可以承担的责任：虽然该过程主要是由青年倡导者设计的，但年轻参与者显然在内容和结果方面最具影响力。

文章阐明了所有参与模式均强调的重要要求：参与应导致可见的结果，这对于年轻的参与者以及城市发展都是有意义的。当然，参与活动可以采取多种形式，而且参与的多种成果和参与形式会相互影响：这既包括临时的建设，也包括对建筑环境的永久更改，或者从单一事件发展成为政策文件。关于参与的结果和影响，同样重要的是，规划专业人员从一开始和整个过程透明沟通已确定选择哪种参与模式、年轻人拥有何种代理机构以及参与过程可能产生什么影响。这样可以避免使参与者期望落空的结果。

尽管目前距离形成儿童和青少年参与的参与式规划文化还有很长的路要走，但是在德国可以找到许多参与式城市发展的最佳实践的例子。这些正面的例子再次强调了年轻人参与城市发展的重要性。因为它们表明，当年轻人参与进来时，对城市空间的新解释可以进入辩论过程，不寻常的思想可以渗入公共空间的设计中，可以申请发起拨款程序来实现以人民为本的邻里发展(参见 Million & Heinrich 即将发表的文章)。

致谢

在本文中，笔者主要借鉴非营利组织“青年建筑城”的工作。笔者在这项工作有双重角色：

一方面,笔者本身就是该组织的成员;另一方面,笔者与柏林工业大学的同事一起研究了"青年建筑城"的工作。毋庸置疑,不仅是笔者的个人经历,而且还有"青年建筑城"和柏林工业大学的同事的观察和思考,都被纳入了这篇文章中。因此,笔者要感谢所有同事,尤其是安德烈·本泽(Andrea Benze),拉尔夫·弗莱肯斯坦(Ralf Fleckenstein),克里斯蒂娜·吉米内斯·马特森(Christina Jiménez Mattson),安吉拉·米利恩(Angela Million),安克·施密特(Anke Schmidt),祖扎娜·塔巴奇科瓦(Zuzana Tabačková)和乌尔斯·沃尔特(Urs Walter)。

参考文献

[1] Arnstein S, 1969. A ladder of citizen participation[J]. Journal of the American Institute of Planners, 35(4): 216-224.

[2] Ataol Ö, Krishnamurthy S, et al, 2019. Children's participation in urban planning and design: A Systematic Review[J]. Children, Youth and Environments, 29(2): 27-47.

[3] Edelhoff S, Fleckenstein R, Grotkamp B, et al, 2019. Handbuch der baukulturellen Bildung (Engl.: Handbook to built environment education)[M]. Berlin: Universitätsverlag der TU Berlin.

[4] Derr V, Chawla L, Mintzer M, 2018. Placemaking with children and youth: Participatory practices for planning sustainable communities[M]. New York: New Village Press.

[5] Hart R, 1992. Children's participation: From tokenism to citizenship[M]. Florence: UNICEF International Child Development Centre.

[6] Hart R, 2008. Stepping back from "The ladder": Reflections on a model of participatory work with children[M]//Reid A, Jensen B, Nikel J, et al. Participation and learning: perspectives on education and the environment, health and sustainability. New York: Springer: 19-31.

[7] Heinrich A, Million A, 2016. Young people as city builders: Youth participation in German municipalities [J]. disP—The Planning Review, 52(1): 56-71.

[8] JAS—Jugend Architektur Stadt e. V., Edelhoff S, Fleckenstein R, et al., 2019. Hingucker 2013-2017: Partizipatives Bauen und Gestalten mit Kindern und Jugendlichen (Engl.: Eyecatcher 2013-2017. Participatory building and designing with children and youth)[M]. JAS e. V.: Essen.

[9] Jupp E, 2008. The feeling of participation: Everyday spaces and urban change[J]. Geoforum, 39 (1): 331-343.

[10] Million A, Heinrich A, 2020 (forthcoming). Kinder-und jugendgerechte Stadtentwicklung. Bildungs-und Experimentierräume für junge Stadtgestalter (Engl.: Child- and youth-oriented urban development. Educational and experimental spaces for young urban designers)[R]//Berliner Beirat für Familienfragen. Berlin: Berliner Familienbericht.

[11] Percy-Smith B, 2010. Councils, consultations and community: Rethinking the spaces for children and young people's participation[J]. Children's Geographies, 8(2):107-122.

Young urban designers: Involving children and youth in urban development in Germany

Anna Juliane Heinrich

Urban planning should be a people-oriented, communicative process in which citizens participate—this attitude has become a matter of course in Germany and many other countries all over the world since the 1970s. Initially, the discourse and practice of communicative planning focused on formats of "being participated"—namely options for citizens to have a say in decision-making that were opened up and moderated by politics and administration. The underlying assumption of a bipolar relationship between the public and the private sphere, however, has dissolved in favor of a much more plural understanding of urban development. Urban development is now understood as interaction between different logics and interests of politics, administration, economy, civil society and intermediary stakeholders.

Taking up and developing this debate on the plurality of urban planning and its stakeholders, I will argue that people-oriented urban development should not remain limited to approaches that focus on urban planners as professionals considering "the people" and designing for people. Although "the people" should of course be put at the heart of planning efforts, this approach falls short. People-oriented design should embrace approaches that actively involve "the people" in urban development. Even more so, I argue that urban development is always also shaped by the citizens and that this should be acknowledged, appreciated and fostered by planning professionals and decision-makers. Undoubtedly, this broad understanding of urban development poses great challenges for urban planners and requires a rethinking that will transform our understanding of urban planning and the roles and responsibilities of the planning profession.

My concern in this contribution is young people and their role in urban development processes. In the case of Germany, children have been understood as a target group of planning for several decades; especially in the context of the development of family-friendly residential areas, they are considered by planning professionals. Adolescents have been increasingly involved in planning processes only in the last fifteen years; however, right from the start, they have mostly been understood as participants to be actively involved in planning processes. In recent years, they are increasingly understood as stakeholders independently shaping urban development.

In the following, I will distinguish three different modes of young people's participation in urban development: state-led participation, youth advocate-led involvement and youth-led projects. This shall broaden our understanding of young people's involvement and open up a spectrum of opportunities for citizen participation and co-production in different contexts. Thereafter, I will focus on youth advocate-led participation—an approach to urban development which might involve state agencies but is in the responsibility of youth advocates as intermediary stakeholders and which motivates and empowers young people to act as urban designers. As an example of such a youth advocate, I will introduce the German non-profit organization YOUTH ARCHITECTURE CITY. The background to this is for one thing that I am myself an active member of this organization and for another that together with colleagues I have done research on participation processes that YOUTH ARCHITECTURE CITY conducted. In order to inspire participation processes, I will describe how the non-profit organisation designs participation along with four phases: exploring, envisioning, designing and communicating. Following this, I will provide insights into one practical example of young people's involvement in the redevelopment of a playground. In this, I will reflect on some success factors and obstacles of young people's participation. In conclusion, I will briefly touch upon urban planning professional's consequences deriving from the different modes of young people's participation.

Different modes of participation: State-led, youth advocate-led and youth-led

In 1992 Roger Hart set a milestone regarding young people's participation: Adapting Sherry Arnstein's (1969) seminal ladder of participation to the involvement of young people, Hart (1992: 8-14) for one thing exposed modes of non-participation that range from manipulation to tokenism. For another thing, he differentiated degrees of participation, ranging from young people being "assigned but informed" to "adult-initiated, shared decisions with children" to "child-initiated, shared decisions with adults" (Hart 1992: 8). These degrees describe differing grades of young people's agency and are in of themselves equally important and need to suit the respective context (cf. Hart 2008: 23).

Despite Hart's eye-opener for a broad variety of degrees of participation and the resulting sensitivity for differing degrees of agency, a rather one-dimensional understanding of participation kept dominating participatory practices for a long time. Young people's participation in processes of urban development was still primarily understood as activities initiated and executed by the state. Procedures and methods remained close to classical formats of democratic opinion formation (cf. Jupp 2008; Percy-Smith 2010; Heinrich & Million 2016).

However, throughout the last fifteen years, a discourse has developed that fosters a broader understanding of participation (for a systematic literature review on children's participation in planning see Ataol et al. 2019). Jupp (2008: 332) criticises that "participation in policy frameworks is often understood to involve quite specific activities and spaces, for example attending meetings in town halls." She counters this, arguing "that the work of small-scale community groups can provide a powerful basis for the engagement, and empowerment of local people, in ways that might include, but certainly not be limited to, such conventional forums of direct interaction with the local state" (Jupp 2008: 332). Accordingly, Jupp points out that "it may be necessary to reconsider what might constitute 'participation'" (Jupp 2008: 333) and in line with this suggests to differentiate between different "modes of participation" (Jupp 2008: 334). Besides "state-led initiatives" (Jupp 2008: 334), "community-led" (Jupp 2008: 341) activities are also to be understood as participation.

Percy-Smith (2010) advances this discourse and explicitly addresses the participation of young people. He states that "we need to move away from the current emphasis on participation in formal, institutionalised public decision-making processes and instead focus more on the multiplicity of ways which people act, contribute to and realise their own sense of agency in everyday life contexts" (Percy-Smith 2010: 119f.). In concrete terms young participants should have "the right to exercise power over the agenda and process in participatory initiatives" (Percy-Smith 2010: 110). He advocates an understanding of participation, which also embraces all sorts of informal, everyday practices of active citizenship (cf. Percy-Smith 2010: 109f.).

Building upon this discourse, Heinrich & Million (2016) distinguish three modes of youth participation: state-led participation, youth advocate-led involvement and youth-led projects. State-led youth participation is understood as a process initiated and managed by a state agency which involves young people in planning processes. The topics, spatial setting, procedure and methods are primarily chosen by the state agency. This top-down participation offers several advantages: the direct link in administration and politics ensures that the voices of young people are heard by decision-makers. This can be a fruitful basis for the implementation of ideas. Furthermore, state-led participation addresses topics and scale levels that are usually not covered by the other two modes: for example the scale levels of the whole city or even region. Planning tasks and documents with a high degree of abstraction and long planning horizons are tackled. Obstacles are, however, that the topic may be completely detached from young people's daily lives and their interests. Recurring problems are the activation and motivation of young people for such participation and the communication between a state agency and youth (cf. Heinrich & Million 2016: 61f. and 68).

Youth advocates, again, "are stakeholders who foster youth participation either independently or on behalf of a commissioning state agency" (Heinrich & Million 2016: 57). A youth advocate could for example be a non-profit organisation or a private urban planning office. Characteristic is that—despite a state agency's mandate—youth advocates clearly act partial towards the young people they involve. Since they would usually have experience in involving young people, they have a "closeness to adolescents and their interests, needs, forms of communication, modes of practice and culture" (Heinrich & Million 2016: 64), which results in meaningful involvement. While this mode is rather characterised as top-down, youth advocates prove to be very sensitive towards the target group, for example being more successful in addressing and activating young people (cf. Heinrich & Million 2016: 64 and 68).

The third mode of young people's participation in urban development is youth-led projects. These embrace "any form of activity and commitment of young people, which somehow develops a meaning or relevance for the urban environment and its citizens" (Heinrich/Million 2016: 68f.). Youth-led projects are bottom-up initiatives characterised by young people acting independently and being in charge of their activities. Young people involve in the issues that they are interested in and that concern them. With their initiatives, they mostly focus on a micro-scale, for example reusing or redesigning single plots, buildings or plazas. They favour short-to medium-term activities with a "hands-on" character. All of this is highly motivating for young people and they act with great commitment. Our research shows that "youth initiatives can make a valuable contribution to neighborhood development, to on-going planning processes and that they can inspire and qualify state-led youth participation" (Heinrich & Million 2016: 63). Usually, these projects are not linked to ongoing planning processes by the city administration. However, it can be worthwhile for state agencies to trace such projects and either support them or involve them in ongoing transformation processes. Yet, these initiatives sometimes perceive such advances as undesirable interference (cf. Heinrich & Million 2016: 63-69).

It is understood, that the described three modes of youth participation are not to be understood as a strict three-category scheme but rather different spheres, which overlap, diffuse and blend into each other (Heinrich/Million 2016: 69). Analysing different practices, we can identify hybrid forms, which meet characteristics of two or even all three modes of participation.

YOUTH ARCHITECTURE CITY: A non-profit organisation as youth advocate

One such youth advocate in Germany is the non-profit organisation YOUTH ARCHITECTURE CITY (German: Jugend Architektur Stadt e. V.; http://www.jugend-architektur-stadt.de/english). The aim of the organisation is to foster young people's participation in processes of planning and building as well as to promote built environment education for children and youth. With its activities, YOUTH ARCHITECTURE CITY invites young people to engage with their environment. Children and youth are encouraged and enabled to perceive architecture, design, neighbourhoods, public spaces, landscapes, cities and regions with all their senses, to (re)discover and shape these spaces. Since the foundation in 2005, the members of YOUTH ARCHITECTURE CITY have tested and developed a broad repertoire of methods for young people's participation and built environment education (see Edelhoff et al. 2019).

Conceptualizing processes of young people's participation: explore, envision, design, communicate

It goes without saying, that each and every process of young people's participation in urban development is unique. Each process is characterized amongst other things by the

— pursued aims (e.g. regarding different fields of action, groups of stakeholders),
— topics (e.g. mobility, meeting places),
— spatial setting (e.g. a plaza, neighbourhood),
— stakeholders involved (e.g. state officials, business people),
— target groups (e.g. an age group, users of a specific place),
— modes of participation (e.g. youth-led, state-led),
— applied methods (for broad spectrum see e.g. Derr et al. 2018),
— availability of resources (e.g. staff, financial resources),
— time frame (e.g. short-term project, longer-term process).

All of these aspects need to be explicitly addressed and clarified either before the start or throughout a participation process because they shape the process. In view of these many factors influencing the process, it is helpful to have an orientation framework that helps to structure a participation process. Therefore, participation processes organised by YOUTH ARCHITECTURE CITY are usually conceptualised along with four phases: (1) explore, (2) envision, (3) design, (4) communicate.

Exploring an environment is usually the first step in a participation process. All participants will be encouraged and enabled to get to know the setting that the participation process refers to. This is especially exciting (but also challenging) in settings in which the young participants are well versed. Methods should be chosen, which help the participants to open their eyes to what is overlooked in everyday life. Furthermore, methods can be chosen that open up new perspectives on a setting. Besides these approaches focusing on the perception of a setting, the phase of exploring also includes the evaluation of qualities. The young participants should appreciate what already contributes to a good quality of living and identify problems. In the course of that, they will need to explore their own needs and their demands on a place in order to establish valuation standards. In order to sensitize the participants for the balancing of different interests in urban development, it also makes sense to motivate young people to explore other stakeholders' needs.

The acquired sensibility for the respective setting is the basis for the second phase: envisioning. In essence, this phase is about understanding, elaborating and articulating what change is needed according to the participants' perspective. As a result of the previous phase of exploration, the participants have a spectrum of potentials which they can build upon and further develop with their ideas. The problems should be addressed by their visions and, if possible, resolved. The aim is to work out alternative solutions and futures for a setting.

The third phase, the phase of designing, includes a decision for one or several of the envisioned ideas. It goes without saying that decision-making needs to be based on a fair and transparent participatory process. Once a decision has been made, the vision usually needs to be translated into concrete measures. Further elaboration is needed to make the vision feasible. What this process of designing actually looks like, depends very much on the nature of the whole participation process (see above: pursued aims, topics, etc.).

The concluding phase is the phase of communication. Undoubtedly, internal communication is of the utmost importance throughout the whole process of participation. However, this phase describes external communication. The young participants' results from the first three phases shall be presented to a wider audience. Different strategies and aims can be pursued with this: Young people's evaluations of their environment as well as their visions for the future development can be communicated addressing the public. This can help to raise awareness in society of the needs of young people. Moreover, especially adolescents are a group that is often burdened with prejudices; presenting their visions can help to reduce these prejudices. However, the core concern of this phase of communication is usually to address decision-makers and to ensure that the participation process has an impact on urban development.

It goes without saying, that these four phases are to be understood as an ideal type of model. Within a participatory process, the phases sure overlap or merge seamlessly. However, working along with these four phases has proven itself: The model serves as a framework which gives structure and guidance without overdetermining.

Young urban designers planning and implementing the redesign of a playground in Berlin, Germany

In 2014 and 2015, YOUTH ARCHITECTURE CITY involved young people in planning and implementing the redesign of a playground.* (Foodsnote) The setting of this renewal project was a rundown areal of a playground in a deprived neighbourhood in Berlin, Germany. The playground is of utmost importance for the young people living in the neighbourhood, especially because it is located between two well-frequented

---

* *The following is based on (1) the project documentation (JAS e. V. 2019: 74-81), (2) the project website (http://www.hingucker-jas.de/index.php/berlin) as well as interviews and discussions with the involved members of the non-profit organization which were carried out as part of accompanying research.*

institutions: a children's leisure centre and a youth club.

The first year of the process was dedicated to children's and youth's participation in the planning of the redesign. A dialogue between young users of the space and the landscape planners in charge of the redesign was to be initiated. The second year served to involve young people in the concrete design as well as the building process of seating furniture for the playground. Both project years were conceptualized along with the four work phases of YOUTH ARCHITECTURE CITY and will be discussed accordingly. However, the first year focused more on exploring and envisioning; while the second year emphasised the phase of designing and the actual realisation of young people's ideas.

The whole process started with an exploration of the playground. On-site young people engaged with the current conditions and their demands for the future development of the areal. In order to record their results, the participants took photographs and designed postcards pointedly stating the results of their evaluation. This served to learn more about young people's perception of the playground and to generate a product with which to communicate this. As a second step young people were invited to learn more about the needs and wishes of all (potential) users of the playground. For this, they interviewed people living in the neighborhood and created short video clips from it (Fig.1).

Elaborating on users' demands and ideas for the redevelopment, the exploration merged smoothly with the phase of envisioning. Pushing ahead with this phase, YOUTH ARCHITECTURE CITY created a simple participatory game for the envisioning and negotiation of different futures of the playground. Based on the young people's postcards and videos, a broad range of uses that the young people miss in their neighborhood was identified. The aim of the game was to carve out the importance of these uses for the young people. Divided into age groups the young people received an equally distributed share of building blocks and could use these to vote for the displayed uses. In the course of the game, these building blocks piled up on the playing field to form towers that displayed the participants' priorities. However, since the availability of space (and money) limits how much could be realized on the playground, the crux of the game was that the young people had to forge alliances to push through their most urgent desires. As a result of playing this game for an afternoon, the young people came up with recommendations for the redesign (Fig.2 & Fig.3).

Two important requirements of the phases of exploration and envisioning can be identified from this project: First, youth advocates should deliberately choose methods that address a variety of modes of communication. In the example photography, postcard design, filming, oral discussions and the atmosphere of the game were chosen. Offering participants different ways to engage in the process, fosters inclusive participation. Different participants will feel comfortable with different modes of communication. Therefore, a broad spectrum should be covered, including verbal (e.g. oral presentation, storytelling, interviewing, writing), visual (e. g. photography, filming, sketching, collaging) and also haptic and atmospheric (e. g. modeling, gaming) means of communication.

Second, youth advocates created two moments in the process, where participants were asked to engage with the needs and demands of others: young people interviewed other users and played the game they negotiated with each other. Letting young participants discover other stakeholders' needs is worthwhile because it gives the young participants a realistic idea of how urban planning always has to meet different needs. It makes planning processes transparent for young people and forms a basis for consensus and cooperation.

The phase of designing was only realized to some extent. In provoking prioritization by playing the participatory game, aspects of the phase of designing were addressed. However, as will be shown subsequently, designing was the focal point of the second year of this process.

The phase of communication was relevant in three respects: First, from their video clips, the young people developed a video message summarizing their demands for the landscape planners in charge of the redesign. Second, the young people's recommendations for the redesign were forwarded to the landscape planners. This transfer of results was organized by the youth advocates who represented the interests of the young participants. Third, the young people's recommendations were printed on large-format posters and displayed on a firewall just next to the playground.

These facets of communication hint at an aspect which is relevant regarding the phase of communication in general: Youth advocates need to ensure that their participants' voices are heard. In the example, YOUTH ARCHITECTURE CITY negotiated with the landscape architects to make sure that the participants' ideas had an impact on the redevelopment process. And indeed the young people's ideas were taken up and fundamentally changed the plans that the landscape architects had already prepared before the participation process. So as not to underperform as tokenism, youth advocates must keep an eye on ensuring that the results of participation actually find their way into planning and implementation processes.

Furthermore, it can be worthwhile to address a wider public—as was done with the posters. This can help to sensitize a neighborhood for young people's needs. Displaying young people's positive contributions can also contribute to reducing prejudices towards adolescents.

The second year of this redevelopment process was dedicated to design and realization. In negotiation with the landscape architects in charge, the consensus was reached that one element of the future playground could be designed and built together with the participants. Picking up on the young people's ideas from the phase of envisioning, the desired seating furniture was chosen for this.

Building upon their work of the previous year, the young participants straightaway started with the phase of designing. Developing and displaying their ideas in models, they elaborated on characteristics this seating furniture should have. Of course, the seating furniture needed to be feasible within the given budget and meet all constructional requirements of the building authorities. Therefore, YOUTH ARCHITECTURE CITY translated the young people's design proposals into a ready-to-implement design.

As a next step, all young people built models of the seating furniture on a scale of 1：20. These models already contained all constructional elements that were to be used on the construction site. This step served to simulate the later construction process so that the participants could imagine the realisation processes (Fig.4 & Fig.5).

The highlight of the participation process was the set up of a participatory construction site for the building of the seating furniture. Guided by pedagogically experienced construction professionals, children and adolescents managed to build their wooden seating furniture on their playground. While the construction process and the finished seating furniture already communicated young people's involvement in the neighbourhood, additionally a public opening ceremony was held to celebrate the young people's success and to communicate their valuable contribution to a wider public (Fig.6 & Fig.7).

Some lessons can be drawn from this project example: In the phase of designing, it might be necessary to adapt participants' suggestions so that they are practicable. However, youth advocates need to be very sensible about this so that the young people continue to identify with their results. Therefore, YOUTH ARCHITECTURE CITY invited the young people to build models of the final design. This process allowed the young people to dig deeper into constructional details and to explain why the design looked the way it did. What is more, this can also be understood as built environment education which gives the participants an understanding of processes of designing and building in general.

One real special feature of this participation process was the participatory construction work. The ideal model of YOUTH ARCHITECTURE CITY does not standardly embrace a phase of realisation because the actual implementation of measures of urban development processes can only rarely be performed participatorily. However, it is a factor of success for young people's participation in urban development processes to produce visible outcomes of some sort. In other projects of YOUTH ARCHITECTURE CITY this has taken very different forms: At the end of a participation process, YOUTH ARCHITECTURE CITY often motivates young people to preserve some material reminder of a participation process that can remain visible for them. This might be a poster which will be hanged up in the institution taking part in the process, or this could mean that young people take home the models they built. This shall support young people in keeping in sight what they accomplished and that they can make a difference and change things. Visibility towards a wider public is for example realised by temporary interventions that attract attention and point at young people's demands and ideas, or by public exhibitions of results, or local media coverage. Visibility of results towards decision-makers has in the past for example been a manifesto summarizing young people's claims towards urban development. This manifesto was handed over to politicians in a public event.

Conclusion: Manifold ways to involve young urban designers

Throughout this contribution, I argued that people-oriented urban planning should embrace the involvement of young people as fellow citizens and that this should take manifold forms. Urban planners as professionals should not only plan for young people but plan with young people and appreciate urban development promoted by young people. Since this undoubtedly poses great challenges for urban planners, this article presented different modes of participation which illustrate how young people can be involved as urban designers. Picking up on Roger Hart's ladder of young people's participation, I fostered a stakeholder-centered perspective which focuses on the question of who is in charge of what in different modes of participation. The modes of state-led participation, youth advocate-led involvement and youth-led projects illustrate how urban development is shaped by young people's knowledge, demands and ideas and these modes offer planning professionals different opportunities on how this can possibly be further enabled, fostered, intensified or channeled.

The focus of this article on youth advocate-led involvement and the work of the non-profit organisation YOUTH ARCHITECTURE CITY was supposed to offer insights and practical "how-to" instructions. It became clear what influence the young participants could have and what responsibilities the youth advocate assumed: While the process was primarily designed by the youth advocate, the young participants were clearly most influential regarding contents and outcomes.

A highly important requirement which accounts for all modes of participation was illustrated: Participation should lead to visible results which are meaningful for the young participants as well as for urban development. Of course, this can take many forms and can be an interplay of diverse outcomes and formats of participation: ranging for example from temporary interventions and installations to permanent changes in the built environment or from events to policy documents. Regarding outcomes and impacts of participation, it is also important that planning professionals communicate transparently from the beginning and throughout the process what mode of participation they use, what degree of agency the young people have and what influence the participation process might possibly have. This shall avoid disappointing participants' expectations.

Although, a participatory planning culture that structurally involves children and youth is still a long way off, many examples of best practices for participatory urban development can be found in Germany. These positive examples once again underline the importance of young people's participation in urban development since they show that when young people get involved new interpretations of urban spaces can find their way into debates, unusual ideas can flow into the design of public spaces, appropriation processes can be initiated and people-oriented neighbourhood development can be realised (cf. Million/Heinrich forthcoming).

Acknowledgment

In this paper, I mainly draw on the work of the non-profit organisation YOUTH ARCHITECTURE CITY. I have myself a twofold perspective on this work: For one thing, I am a practitioner of the organisation. For another thing, together with my colleagues at TU Berlin, we research the work of YOUTH ARCHITECTURE CITY. It goes without saying that not only my personal experiences but also observations and reflections from my colleagues from both, YOUTH ARCHITECTURE CITY and TU Berlin, have been incorporated in this contribution. Therefore, I would like to thank all of my colleagues—especially Andrea Benze, Ralf Fleckenstein, Christina Jiménez Mattson, Angela Million, Anke Schmidt, Zuzana Tabačková and Urs Walter.

基于风貌塑造的山地地段城市设计实践研究:重庆案例

Site-based urban design practice for cityscape shaping in mountainous cities: A Chongqing case

杨　震　朱丹妮　陈　瑞　汪　乐
Yang Zhen　Zhu Danni　Chen Rui　Wang Le

摘　要:基于经典城市设计理论,构建了城市风貌5个核心维度:“意象系统”“视觉序列”“形态适应性”“公共空间”“生态肌理”。以山地城市重庆为案例,展示了“以风貌问题为导向——以策略推导总体方案——以要素归纳导控要求”的地段城市设计实践过程,从5个核心维度模拟测评其塑造的风貌结果,分析了城市设计与现行规划管理体系和开发机制的适应方式。由此形成一套具有普适性的城市设计实践方法、评价体系和实施路径,符合塑造山地城市风貌及提升城市治理的需求。
关键词:山地城市;城市设计;城市风貌;城市治理

Abstract: This paper concludes five dimensions of cityscape from tested urban design literatures: the image system, the visual order, the form fitness, the public space, and the ecological pattern. Based on a case study in the mountainous city Chongqing, the paper introduces a site-based urban design process including cityscape-problem analyses, strategy deduction and control-element formulation. The paper measures the simulative effectiveness of cityscape shaping by examining the foregoing five dimensions, and discusses the way accommodating urban design into current urban planning and development mechanism. In all, in the need of cityscape shaping and urban-governance promotion for mountainous cities, the paper presents a universal set of urban-design practical methods, a measurement system and implementation means.
Key words: Mountainous Cities; Urban Design; Cityscape; Urban Governance

1　新时期的城市风貌危机

自中国城市化率突破50%以来,在高速城市化之后“城市风貌的雷同、混乱、缺失”正成为一个日益受到关注的现象，这在众多批判性的语境内被阐释为"千城一面""杂乱拼贴""尺度失调""特色失落"等症状①。在城市设计范畴，"城市风貌"被认为是城市物质形态（包括人工建成环境与自然支撑环境）与社会内涵（包括文化、政治、经济等维度）的同构体：前者是"城市风貌"的外在表征，体现为"貌"（外貌、形貌）；后者是"城市风貌"的隐性特质，体现为"风"（风格、气质）[1-3]。一般认为，城市风貌应该体现"在地性集合"（localities），具有鲜明的"可识别性"（legibility）[4-5]。前述种种城市风貌的症状，本质上是由于"在地性异质化或者失落"，外化为"可识别性的趋同与混乱"[6-7]。这种"城市风貌危机"不是中国城市特有的现象，但中国的城市化速度与尺度在很大程度上放大了城市风貌失控的程度，达到"愈演愈烈"的地步，成为后城市化时期城市治理需要着力解决的问题[8-9]。而城市设计作为一种以塑造城市形态和公共空间为本源任务的"公共政策"，被认为是化解城市风貌危机的主要手段之一[10-12]——这在 2013 年中央城镇化工作会议、2015 年中央城市工作会议、2017 年住建部城市设计试点工作座谈会等会议中被反复强调，促成在全国范围内开展以提升城市风貌为重要目标之一的城市设计试点工作②。在此背景下，本文意图探讨在特定地域条件下，如何通过城市设计结合在地性特质，塑造微观地段的城市风貌③，进而尝试归纳在当地条件下具备一定普适性的城市设计路径，实现与城市治理的有效结合。

2 相关理论：城市设计与城市风貌

在经典城市设计理论中，大量的讨论与城市风貌相关。其中，得到广泛应用的是"城市意象"理论（city image）[13]："路径、节点、边界、区域、标志物"这 5 个要素被认为是塑造城市意象的基本语汇，在城市体验者的头脑中构成城市风貌的"认知地图"（cognitive map）。"城镇景观"理论（townscape）[14]则强调基于运动中的人对城市风貌的"视觉体验"，以创造一种从整体到局部互相联系、"可视化"的城市风貌系统。"好的城市形态"理论（good city forms）[15]进一步提出为了塑造良好的风貌，城市形态应首先具备"活性""适用""可达"3 个特征，它们分别指城市形态保护生态活性的能力、城市形态与人类活动特征匹配的程度、城市空间便利获取各种资源的能力（包括自然景观、社区服务等）。"模式语言"（pattern language）[16]、"拼贴城市"

---

① 2014 年 2 月中共中央总书记习近平考察北京城市建设时提出"守住城市灵魂，保持城市个性"，10 月出席文艺工作座谈会时指出"不要搞奇奇怪怪的建筑"；2015 年中央城市工作会议指出"要加强对城市的空间立体性、平面协调性、风貌整体性、文脉延续性等方面的规划和管控"；《2016 年中共中央国务院关于进一步加强城市规划建设管理工作的若干意见》强调，反对城市存在的"建筑贪大、媚洋、求怪、特色缺失和文化传承堪忧"等问题。这些观点均与城市风貌问题相关，高速城市化之后中国城市风貌方面的乱象成为近年来业界与学界的关注重点。

② 2013 年中央城镇化工作会议提出城市建设要"望得见山、看得见水、记得住乡愁"，要"发展有历史记忆、地域特色、民族特点的美丽城镇"，这些要求均直接指涉城市设计的本体范畴；2015 年中央城市工作会议明确提出"要加强城市设计，提倡城市修补"，"留住城市特有的地域环境、文化特色、建筑风格等'基因'"；2017 年住建部城市设计试点工作座谈会指出城市设计工作的重点是"实现对城市格局、风貌和各类空间的精细化管理"，改变"千城一面、缺乏特色"等问题；住建部于 2017 年 6 月颁布《城市设计管理办法》，明确新时期城市设计工作目标是"提高城市建设水平，塑造城市风貌特色"。

③ 住建部的《城市设计管理办法》将城市设计分为"总体城市设计"与"重点地区城市设计"。本文研究的"地段城市设计"从属于后者的范畴，但更聚焦于微观尺度，并与土地开发行为结合得更紧密。

(collage city)[17]、“城市单元”(urban components)[18]等理论则都倡导“循序渐进”和“致密紧凑”的城市风貌构造原则,强调城市风貌是“城市空间范型”(typo-morphology)的有机演化与各种“在地性营造活动”适应性叠合的结果,反对以“批发式”或者“简化主义”态度来营建城市风貌。“场所感”(sense of place)[19]、“场所个性”(identities of place)[20]、“场所营造”(place making)[21]等理论更强调“公共空间”的重要性,认为公共空间是联结“人工环境—自然环境—社区/人”这三者的纽带,是城市风貌塑造过程中的核心要素。上述理论大多聚焦于人工建成环境,对于自然支撑环境对城市风貌的影响涉及不多。而近年来兴起的“生态都市主义”(ecological urbanism)[22]、“景观都市主义”(landscape urbanism)[23]则明确提出自然生态肌理是城市风貌系统中的“设计要件”(imperative design elements),应该以一元化的视角将其同等纳入城市设计进程,甚至在某些情况下应将生态肌理作为“绿色基础设施”来统领城市风貌的塑造[24-25]。

基于上述讨论,笔者尝试归纳出城市风貌的5个核心维度,分别是“意象系统”“视觉序列”“形态适应性”“公共空间”“生态肌理”,构成一个可用以检验微观地段城市风貌的总体框架(图1)。如果这5个维度都能通过城市设计得到恰如其分的凸显并建立相互联系,进而激发城市体验者的清晰认知,则可以认为在较大程度上塑造出了一个整体而明确的城市风貌。

图1 构成城市风貌的5个核心维度
(来源:作者自绘)

3 山地城市:特定地域的风貌问题

山地约占中国陆地面积的67%,山地城市约占全国城镇总数的50%,典型山地城市包括重庆、贵阳、昆明、桂林、香港等。山地城市具备独特的地域特征:坡降大、比高悬殊,适宜建设用地缺乏,城市环境往往紧密依附于山体与水体。这些特征奠定了山地城市特别的空间本底,构成显著的“在地性”;通过适宜的城市设计手段,能够塑造出与平原城市迥然不同的山地城市风貌①。

然而,在高速城市化之后,中国山地城市也面临普遍的风貌危机,主要包括三方面:

(1) 整体意象紊乱和视觉认知体系失序。体现为建成环境对山水空间的持续侵占及蔓延,由于缺乏系统和连贯的城市设计导控,造成空间环境的叠加紊乱,城市意象的关键要素难以得到明确的厘清和解读,进而无法有效建构具备“可识别性”的视觉体系,造成“风貌雷同、千城一面”等现象。

(2) 城市形态与自然环境适应性耦合不足。体现为建成环境对山地地形地貌的本底格局和立体形态采取傲慢对立或盲目改造的方式,造成人工环境与自然环境的肌理冲突与形态抵

① 山地城市塑造良好城市风貌的典范包括捷克的布拉格(Prague)和克鲁姆洛夫(Cesky Krumlov)、西班牙的托莱多(Toledo)、意大利的锡耶纳(Siena)等,这些时常被当作城市设计理论研究的经典案例。

牾，违背“好的城市形态”的一些基本原则，外化为“遮山挡水、尺度失调”等问题。

(3) 生态肌理碎片化及生态社会功能失落。体现为山地城市丰富致密的生态基质面和生态要素(如高地、丘陵、槽谷、滩涂地、消落带、湿地等)没有被有效整合为网络化的“绿色基础设施”，难以成为引导开发建设的“设计要件”；相反，常见的是生态肌理的断裂和破碎、生态异质性与景源特征的丢失与湮灭，同时，大量碎片化的生态空间也未能在微观尺度提供社会功效，沦为消极的“景观点缀”。

在某种程度上，山地城市面临的风貌危机，既是后城市化阶段城市风貌问题的普遍体现，也是特定地域条件下“在地性异质化”的集中与放大(图 2)。从现有城市设计研究看，在微观地段层面，还较为缺乏针对山地城市风貌整体性、多维度关联的研究，也较缺乏基于城市真实开发逻辑、适应城市治理需求的城市设计路径的探讨[26]。基于此，本文聚焦典型山地城市重庆，以一个城市设计实践案例展开分析，试图填补现有的研究空白。

图 2　中国山地城市面临“风貌雷同、千城一面”“遮山挡水、尺度失调”“生态肌理破碎”等普遍的风貌问题
(来源：作者自摄)

4　重庆案例研究

4.1　案例背景

重庆是全国开展城市设计较早、较多的城市之一①。但颇有争议的是，多年以来城市设计主要服务于城市的增量发展，是当地政府“促进空间生产、谋求土地绩效”的一种经济化工具，塑造城市风貌并非其中的核心内容[27]。目前，重庆主城区是全国开发强度和密度最高的区域之一，前述种种风貌问题开始凸显②。而在过去 2～3 年中，重庆仍然存在对增量开发的路径

① 见：《重庆渝中半岛城市形象设计规划控制管理规定》(2004)、《2000—2005 重庆市优秀城市规划设计作品集》《2008—2011 重庆区县规划》《2010—2011 重庆主城区“二环时代”大型聚居区规划设计国际方案征集方案汇编》。

② 重庆总城区面积为 7 026.55 平方千米，建成区面积 1 329.45 平方千米(建成区面积排在全国第二位，仅次于北京)，2016 年城镇化率达到 62.6%，城区总人口 1 337.95 万；由于重庆有大量不能用于建设的山体和水体，城区开发度(建成区面积/总城区面积)低，仅为 0.19，但城市总体容积率为 1.02(北京 0.82，纽约 1.23，东京 0.99)，因此在建成区范围内，开发强度与密度很高。见：张舰，《中外大城市建设用地容积率比较》，《城市问题》2015 年第 4 期。

依赖[①]，但可开发用地的面积逐渐变小，过去以“平方千米”为单位的“造城式”开发已不复出现，更多是在建成环境中的“植入式开发”[②]。

在此形势下，重庆的城市设计亟须从侧重空间利益的挖掘转向基于山地地域特征的城市风貌的重塑，从关注大尺度、跃迁式的“城市域问题”转向关注小尺度、微单元式的“地段域问题”。因此，重庆市规划管理部门于2016—2017年在主城区选择若干依山面江、地势复杂、面积较小的用地，开展“地段城市设计”试点工作。其核心目标是：(1)在微观地段尺度，构建体现重庆城市风貌与山水自然形胜适应性协调的城市设计控制体系；(2)在现行管理框架内，明确城市设计管控边界和管控重点，与上位规划及项目开发有机衔接。

可以看到：一方面，试点工作响应重塑城市风貌的需求，体现出对原有城市设计范式的潜在“反正”，更强调基于当地地域特征的、系统化、普适化的解决思路（“控制体系”）；另一方面，试点工作遵循城市真实的管理和开发逻辑（“上位规划”“项目开发”），体现出对当前城市运转机制的理性延续[③]。2017年，重庆成为全国第二批城市设计试点城市之一，这为该次试点做了政治加持[④]。

4.2 问题与制约

重庆巴南区花溪街道滨江地段是试点项目之一。用地面积约42公顷，西临长江，地势西低东高，最大高差约62米，是典型的临江依山地形。用地现状呈现出“意象要素模糊、视觉序列紊乱、城市形态适应性不足、公共空间消极、生态肌理破碎”等典型风貌问题；而现有的控制性详细规划采取二维导向的用地布局方式，在建筑高度控制、组群塑造、公共空间等方面没有明确导控要求，对城市风貌的指引是缺位的(图3)。

同时，用地已被政府平台公司收储，控规的核心指标体系无法改变，具体体现在：(1)开发用地总面积不能减少（即不能将开发用地转化为公共绿地、广场用地）；(2)商务商业用地与居住用地不能互相转移平衡（因为这两类用地的计价体系不同）；(3)开发总建筑面积不能减少（即容积率不能降低，因为它与土地出让金收益相关）；(4)商务商业总建筑面积与居住总建筑面积不能互相转移平衡（同样关系到土地计价体系和出让金收益）[⑤]。

---

① 见：https://cq.focus.cn/zixun/f0edcf855dd55365.html，搜狐焦点。相关数据显示：重庆市2016年推出的建设用地面积相比2015年下降25%，但成交金额上升6.04%，成交楼面地价上升25%，呈现供应减少、需求活跃的特征。

② 见：http://www.cqgtfw.gov.cn/scxx/tdgyjg/index_2.htm，重庆市国土资源和房屋管理局网站土地出让结果。以2017年6月为例，重庆市主城区共出让二类居住用地16宗，其中最大宗地面积23.7公顷，最小宗地面积2.5公顷，平均宗地面积12.7公顷。

③ 尽管有学者建议：可考虑用城市设计取代控规（见：金广君，《控制性详细规划与城市设计》，《西部人居环境学刊》2017年第4期），但在政府层面，仍把城市设计视为对控规的有效补充，两者并行从属于现有的法定规划体系。2017年住建部城市设计试点工作座谈会即指出“城市设计是弥补城市规划体系中缺乏‘设计城市’的环节”，而非取代控规。《城市设计管理办法》中也规定：“重点地区城市设计的内容和要求应当纳入控制性详细规划，并落实到控制性详细规划的相关指标中”“重点地区的控制性详细规划未体现城市设计内容和要求的，应当及时修改完善”。重庆市2016—2017年开展的大部分城市设计试点工作中，均强调城市设计要实现与上位规划的有效衔接，实现城市设计与控规的“双控”机制。参见：重庆市规划局《重庆巴南区花溪街道滨江重要地段城市设计征集文件》(2016年7月5日)。

④ 参见：住建部《关于将上海等37个城市列为第二批城市设计试点城市的通知》(2017年7月25日)。

⑤ 关于核心开发指标不能改变的要求，明确写在城市设计的任务书中。参见：重庆市规划局《重庆巴南区花溪街道滨江重要地段城市设计征集文件》(2016年7月5日)。

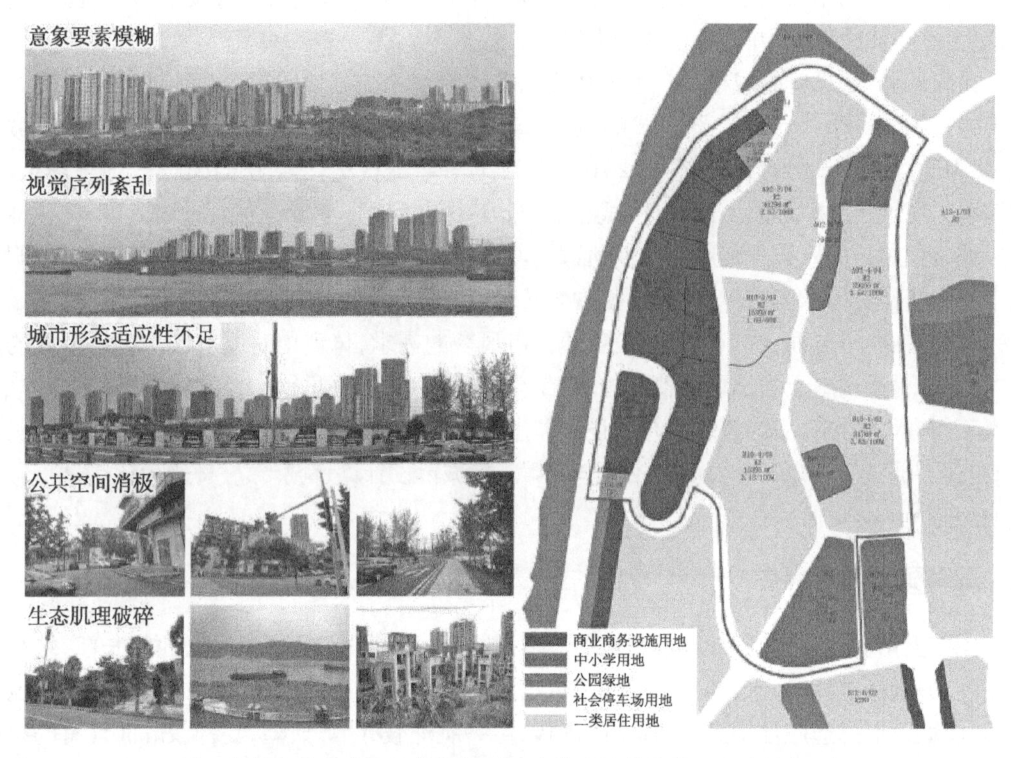

图3　案例区域的现状风貌问题(左)及缺乏风貌导控要求的控规(右)
(来源:作者自制)

4.3　城市设计策略

上述条件体现了当前城市真实的管理与开发机制对城市设计的刚性制约,意味着无法通过减少开发量这种"理想方式"来调节空间形态并塑造城市风貌。基于此,设计团队从"用地结构布局优化、建筑群体关系塑造、公共空间体系建构"等方面着手,制定了6个城市设计策略(图4):

(1) 策略1:"显空"。在用地内预控景观窗口,达到留白透气、连通山水。

(2) 策略2:"透绿"。整合生态肌理,打造公共开放的绿色廊道,实现增绿、活绿。

(3) 策略3:"塑群"。细分建筑群体组合与高度层次,做到疏密有致、形态鲜明。

(4) 策略4:"联岸"。构建多地台的滨江商务商业形态,连通滨江岸线与用地腹部。

(5) 策略5:"聚人"。提升空间的"活性、适用性、可达性",营造积极公共域,凝聚人气。

(6) 策略6:"望江"。增加观江节点,全面提高居民对江景的视线可达性。

这6个策略以前述具体的风貌问题为导向,遵循控规的基本指标体系,同时立足于城市设计本体方法(城市形态和公共空间),体现出对山地城市"在地性"的紧密关联(如生态肌理、多维地形、滨江岸线、山水视窗与廊道、居民行为与山地形态的适应性等)。在此基础上,推导出城市设计总体方案(图5)。

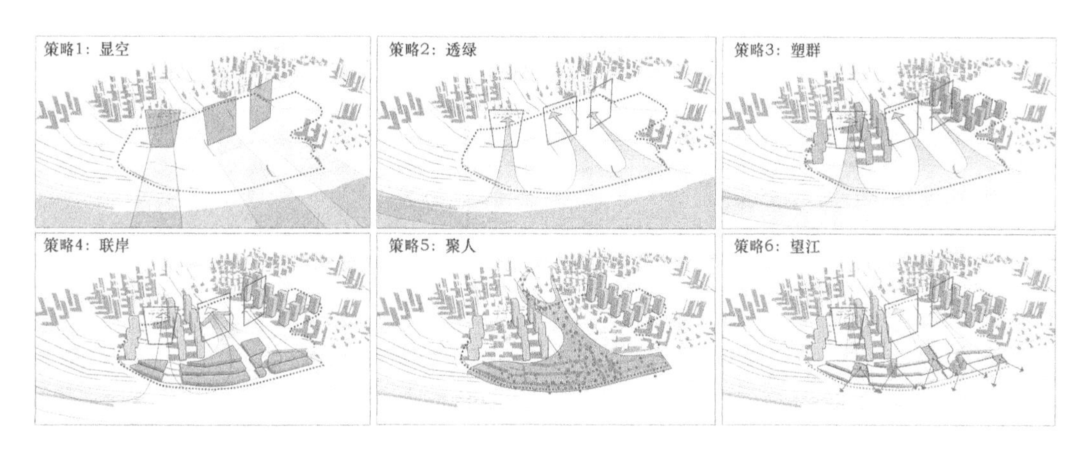

图 4　6 个城市设计策略示意
(来源:作者自绘)

图 5　由城市设计策略推导出的城市设计总体方案(左:沿江鸟瞰图;右:腹地鸟瞰图)
(来源:作者自绘)

4.4　城市设计解析及控制要素

在上述 6 个策略指引下,进一步解析出 7 个城市设计要素:

(1) 要素 1:“山水视窗”。在用地内划定 2 个控高区,降低控高区内居住用地容积率,提高建筑密度,从而降低平均建筑高度(控制在 24～50 米之间),由此形成 2 个显著的“山水视窗”(视窗宽度在 150～300 米),均衡分布于用地北部及中部。这有助于形成沿江全视域范围内的视景窗口,达到滨江空间界面“显空、留白、透气”的效果,与当前沿江建筑密集排布的布局方式形成显著差异(图 6)。

(2) 要素 2:“江山城岭”。将“山水视窗”控高区内降低的容积率转移到其他居住用地,形成 3 个高强度开发区域。在此基础上,对各用地进行高度控制细分:沿江头排建筑限高 75 米,中部建筑限高 100 米,最后一排建筑高度允许达到 120 米。由此在不减少开发总量的前提下,形成“从滨江向腹地逐次上升”的天际线变化,塑造出 3 座“江山城岭”。这种高层建筑纵向排列的布局方式,不同于将住宅沿江横向均质摆放的惯常做法,与用地西低东高的地形适应度更好,同时有利于提升腹部区域居住建筑观江视线的均好性(图 7)。

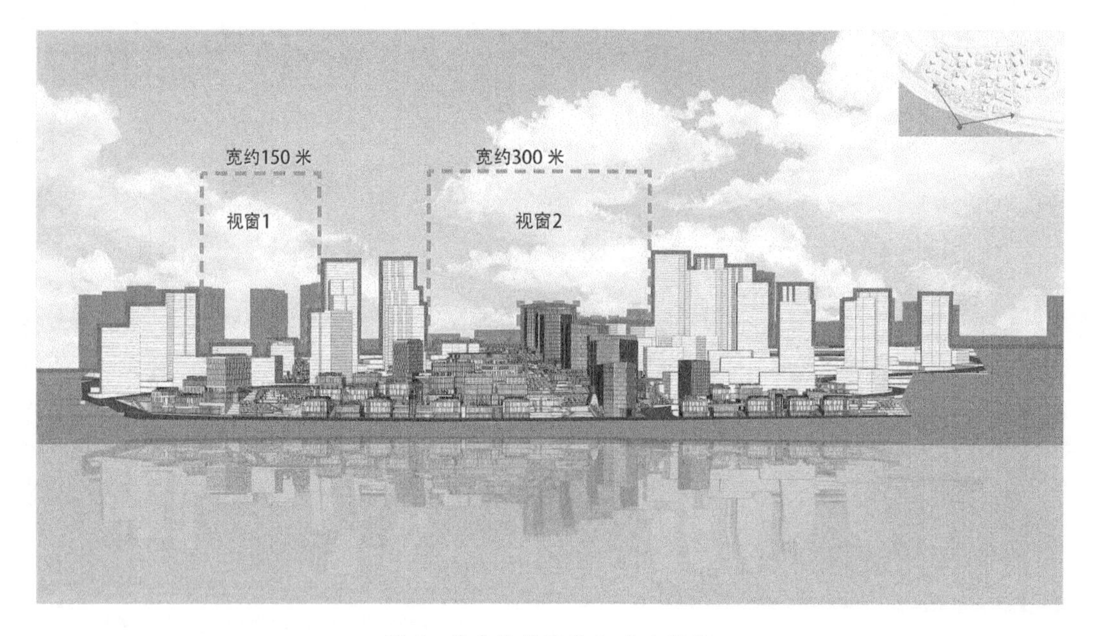

图6　城市设计要素1：山水视窗
（来源：作者自绘）

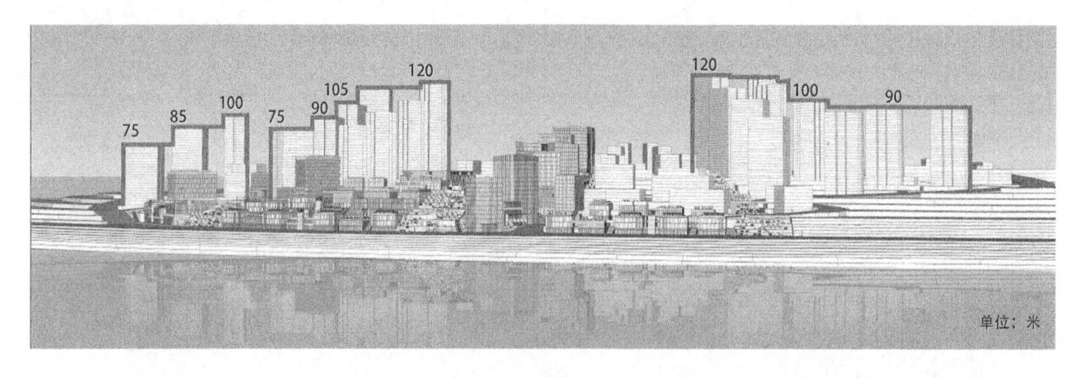

图7　城市设计要素2：江山城岭
（来源：作者自绘）

(3) 要素3："叠翠绿脉"。将原控规分散布局的3块公共绿地等量整合为1条贯穿用地中部、直达滨江的带状公园，赋予其运动休闲功能，以提升公共绿地的空间活性及服务社区的效能。此外，在北部和南部的开发区域，控制2条贯穿式的"私属化公共绿道"，其用地权属及管理职责属于开发者(计入各用地的绿地率指标，不损害核心开发权益)，但其开口位置及界面形态均须遵从城市设计要求。3条"叠翠绿脉"构成用地内完整的生态框架，同时有效限定住区边界，避免出现大尺度封闭住区(图8)。

(4) 要素4："多维江岸"。在3条"叠翠绿脉"临滨江处，嵌套3块商业用地，充分结合临江用地高差，规划分台式商业，提供多首层商业内街和多点式独栋商业组群，形成1条层叠、立体、互联互通的"多维江岸"，避免沿江开发中常见的商铺各自为政、缺乏空间层次的单调形态(图9)。

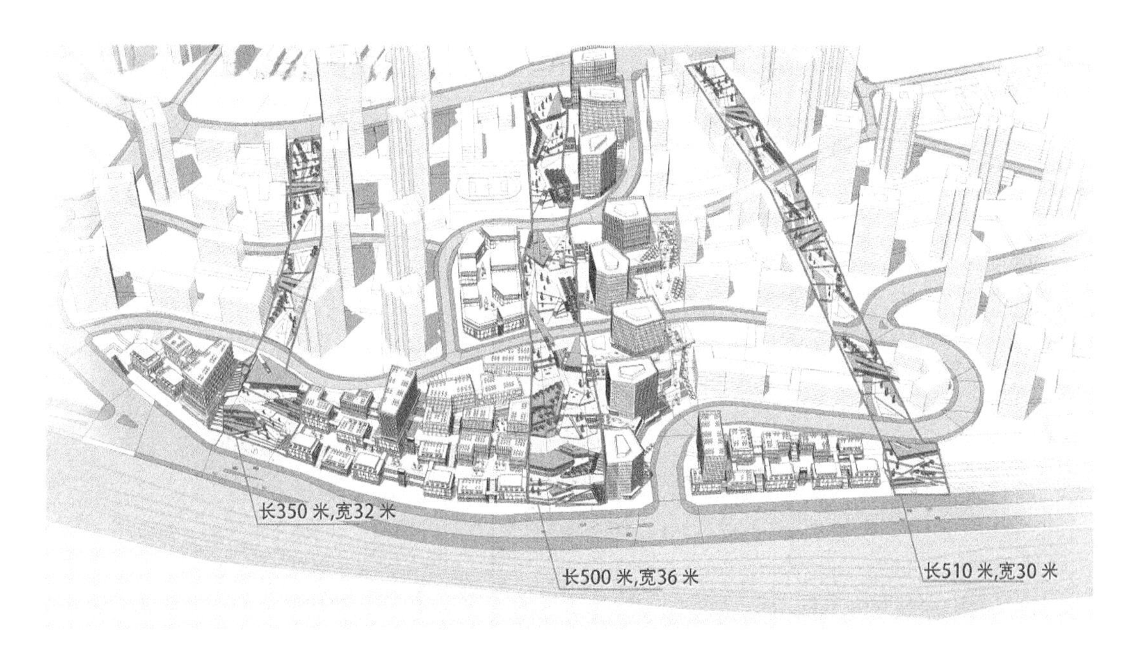

图8　城市设计要素3:叠翠绿脉
(来源:作者自绘)

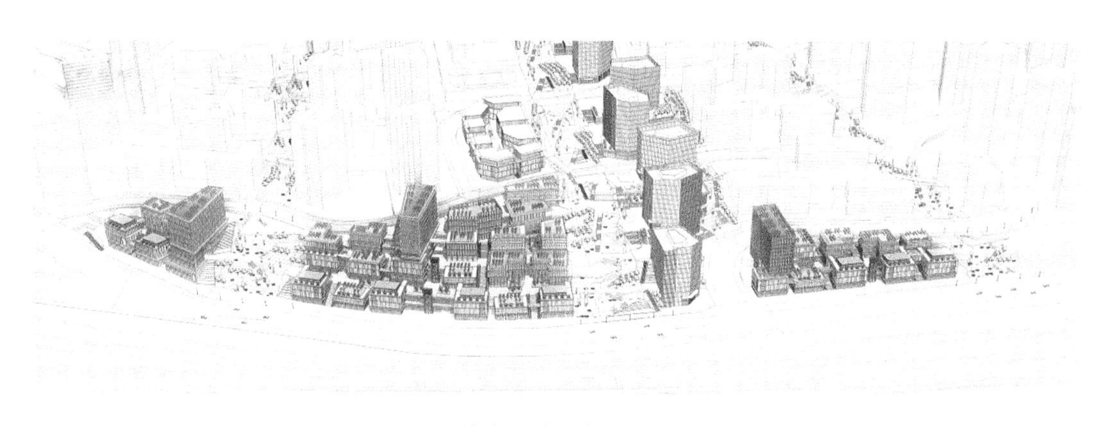

图9　城市设计要素4:多维江岸
(来源:作者自绘)

(5) 要素5:“峡谷金街”。在南北向的“多维江岸”中部,延伸布局1条东西向的“峡谷金街”,建立起T字形的商业空间结构,突破控规中商务商业“沿江一层皮、内外不连通”的问题(商务商业用地总面积和建筑规模均没有减少,只是随用地的优化调整进行了转移平衡)。此外,“峡谷金街”依山层叠、分级而上的形态与传统山地城市意象构成“空间拓扑”的呼应(图10)。

(6) 要素6:“立体公园”。在“峡谷金街”中布局多栋商务商业楼宇,除控制其建筑组群和整体天际线,城市设计的关注重点还包括各楼宇与分台地形的结合及近人尺度公共空间的塑造。采用山地城市常见的建筑接地方式(架、挑、错、退、叠、悬等),串联各楼宇在不同标高的接

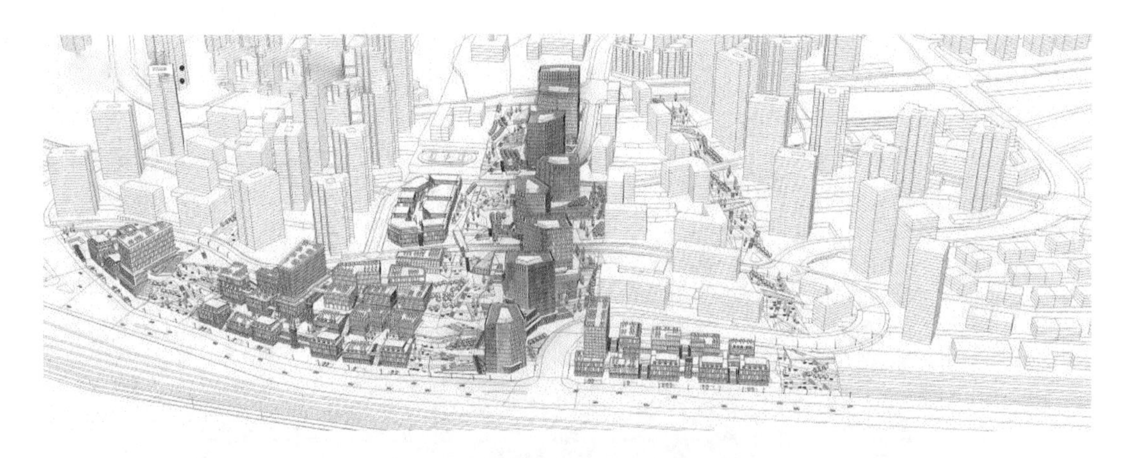

图10　城市设计要素5：峡谷金街
（来源：作者自绘）

地层，形成连续的步行系统及分台绿化，构成总面积约12 000平方米的“立体公园”，为市民提供游山、赏绿、休闲的开放空间(图11)。

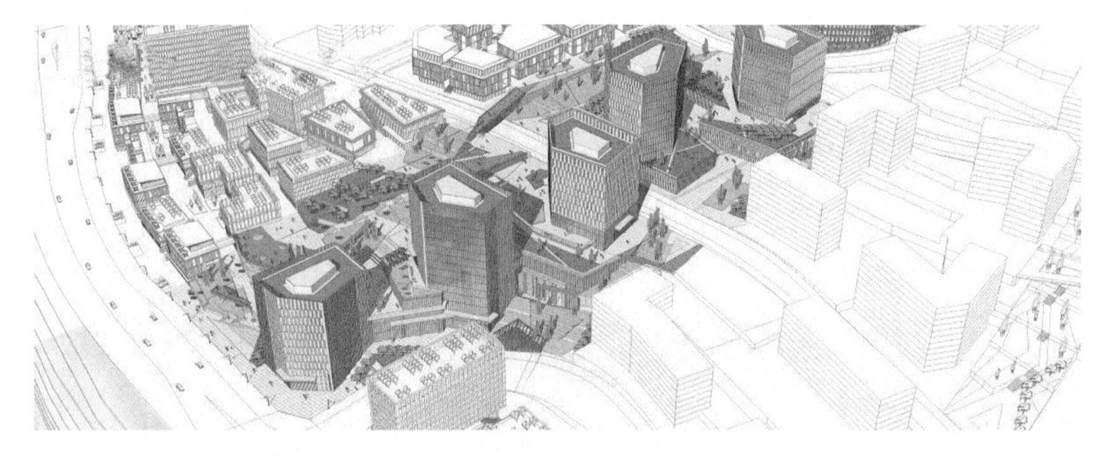

图11　城市设计要素6：立体公园
（来源：作者自绘）

(7) 要素7：“观江阳台”。在临江视线开阔、地势易达处，布点控制5个“观江阳台”，确保市民在多角度能远瞰江景。其中3个“观江阳台”为真正的公共空间(室外开放广场，权属为公共用地)，另2个“观江阳台”则采用“私属化公共空间”的方式，布点在建筑架空层平台处，通过强制性的城市设计导控确保其可用面积和公共可达性(图12)。

上述7个要素可以形成“控制要素引领”(control-element-led)，对城市风貌的核心导控内容进行分项归纳与阐释，制定出规定性或者绩效性的控制要求[28]。这有别于“基于单块用地”(plot-specific)的一般性城市设计控制操作，更符合城市风貌的整体性及系统性特点(体现在城市风貌各个核心维度跨越单块用地，不可分割及互相之间的紧密关联)。

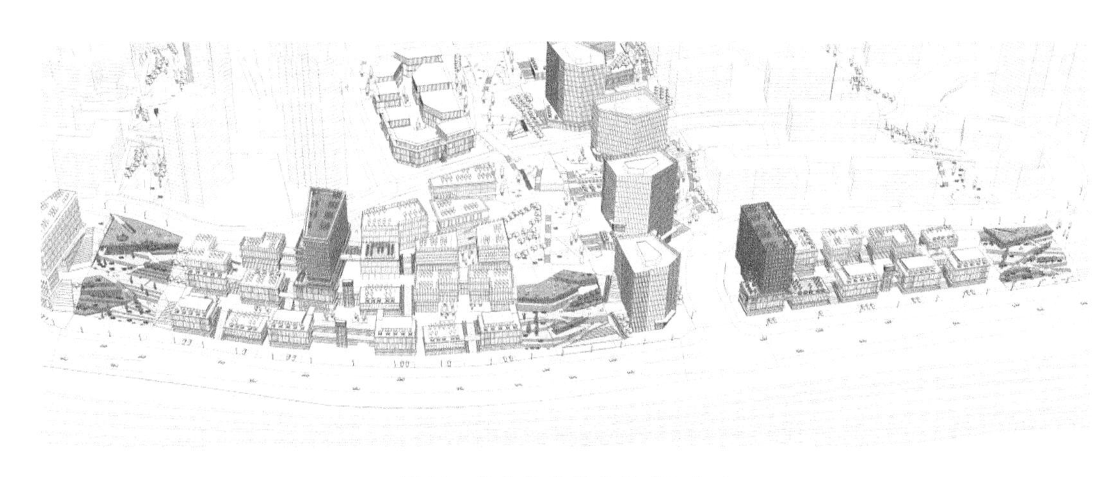

图 12　城市设计要素 7:观江阳台

(来源:作者自绘)

4.5　城市风貌模拟测评

按策略及要素推演出来的城市设计方案是否能构建鲜明的城市风貌?这指涉到城市设计的“实效”。对于还不具备实测条件的项目,可以采取模拟测评的方式,为未来的实测建立框架和基础[29]。笔者依据前述城市风貌的 5 个核心维度,对城市设计成果进行模拟测评。

4.5.1　意象系统

借鉴凯文·林奇(Kevin Lynch)的城市意象研究方法[13],笔者将城市设计方案向 100 名志愿受访者做了展示(客观展示 VR 动画及效果图,设计师避免做过多主观阐述,以免产生诱导),同时辅以简短的访谈,以获取受访者对城市设计方案的意象描述。访谈结果通过 SPSS 分析软件进行归类整理,制作关键词云图(图 13)。

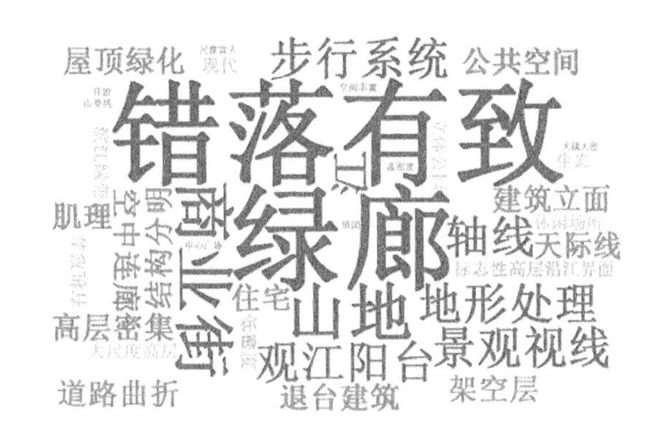

图 13　依据意象调查制作的关键词云图

(来源:作者自绘)

关键词云图分析的主要结果如下所示:

(1) 受访者提及频率最高的关键词是“绿廊”(56%)和“错落有致”(47%)。这主要指涉“叠翠绿脉”及“江山城岭”两个要素,反映“透绿”及“塑群”两个策略。可以推断,这构成受访者认知程度最高的意象。

(2) 提及频率次高的关键词是“山地”(35%)、“商业街”(34%)、“观江阳台”(28%)、“步行系统”(28%)、“景观视线”(27%)、“轴线”(25%)、“地形处理”(24%)。其中“观江阳台”本身即是城市设计要素(反映“望江”策略);“商业街”与“步行系统”主要指涉“多维江岸”及“峡谷金街”(反映“联岸”策略)；“景观视线”及“轴线”仍然与“叠翠绿脉”相关，同时也反映“显空”策略；而“山地”与“地形处理”虽然不直接指涉具体要素，但表明城市设计的空间形态与山地形貌的适应性结合给受访者留下较深印象。

(3) 其他提及频率较高的关键词有“屋顶绿化”(13%)、“公共空间”(12%)、“退台建筑”(12%)、“架空层”(11%)等，较明显地指涉“立体公园”(反映“聚人”策略)。

虽然受访者样本量不大，但关键词的集中度较为明晰，再考虑到大多数受访者不具备设计背景(难以使用“专业语汇”来描绘方案，例如“山水视窗”就较难被受访者明确表述)，因此上述结果对方案整体意象的概括可以认为是基本客观的。

在此基础上，笔者还进一步请受访者使用平板电脑即时绘制“认知地图”。结果显示：大多数受访者(65%)能“默绘”出绿脉、高层建筑天际线、建筑组群等，显示“叠翠绿脉”“江山城岭”“多维江岸”等要素的确给受访者留下了较深的印象(图 14)。

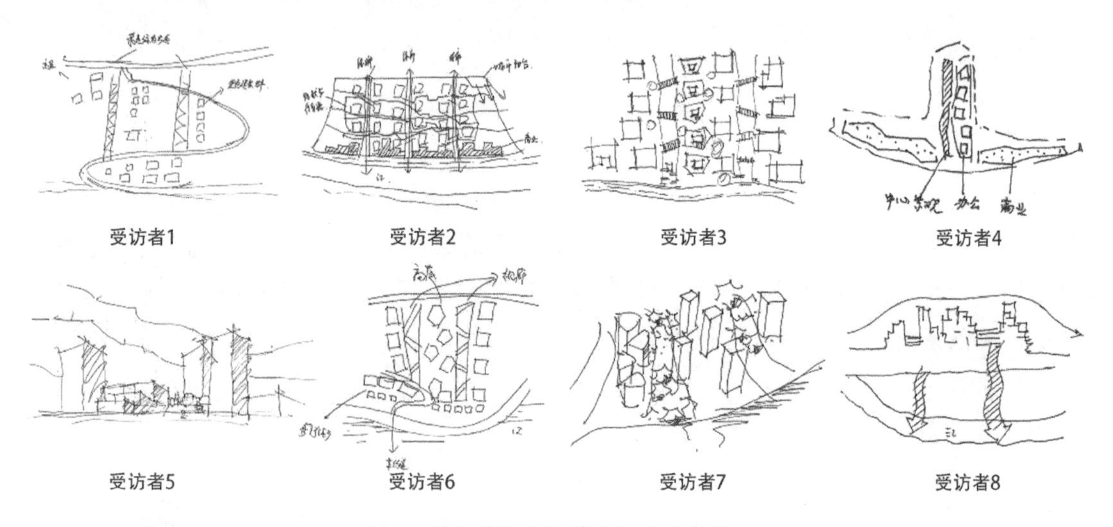

图 14　部分受访者即时绘制的意象认知

(来源：作者收集整理)

4.5.2　视觉序列

参考戈登·库伦(Gorden Cullen)的城镇景观视觉研究方法[14]，结合在建成环境视觉分析中常见的“视域密度”方法[30]，笔者设定 3 个指标参数来反映城市设计方案的沿江形态排列规律及视觉效果：(1)“空域密度”，指在确定范围的沿江展开面中，背景天空面积占沿江界面面积的比例；(2)“实域密度”，指在确定范围的沿江展开面中，实体建筑界面占沿江界面面积的比例；(3)“景物密度”，指在确定范围的沿江展开面中，视域内能明确看到的绿地、山体等自然景物占沿江界面面积的比例。合成分析结果如下(图 15)：

(1) 从用地北部位置分析，未来建成环境的“空域密度”为 31.8%，“实域密度”为 40.8%，“景物密度”为 27.4%。

(2) 从用地中部位置分析，未来建成环境的“空域密度”为 24.2%，“实域密度”为 52.0%，“景物密度”为 23.8%。

(3) 从用地南部位置分析，未来建成环境的“空域密度”为 40.5%，“实域密度”为 32.8%，

"景物密度"为 26.6%。

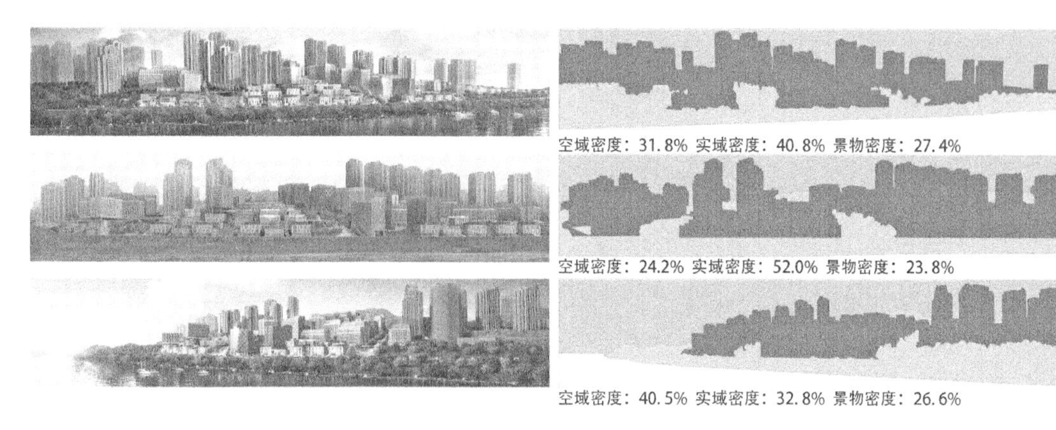

图 15　城市设计方案的视域密度分析(从上到下依次为从用地北部、中部、南部位置分析)

(来源:作者自制)

总体来看,"空域"及"景物"占据沿江界面的较高比例,构成较明确的"空—实—景"形态排列规律,具备较显著的视觉"分形特征"及有机的视觉序列动态变化,与周边建成环境形成明显对比(图 16)。显然,这主要源于"显空""透绿""塑群"的策略指引,以及"山水视窗""叠翠绿脉""江山城岭"等要素的叠加效果。

图 16　现状环境(左上、左下)与城市设计实景合成(右上、右下)对比

(来源:作者自制)

4.5.3　公共空间、生态肌理、形态适应性

由于这 3 个维度具备较紧密的内在关联,并且包含可量化分析指征,如可达性、完整度、适应性[29],因此笔者采用 ArcGIS 系统及 Depthmap 软件做了整体量化模拟测度。主要结果如下:

(1) 公共空间可达性:利用 ArcGIS 网络分析模块中 OD 成本矩阵功能,对开放广场、公共绿地、观江阳台、屋顶平台等主要公共空间进行时间成本可达性分析,发现在对道路行走速度、开放空间吸引度和出行概率赋予不同权重的情况下,开放广场的步行可达时间均在 5 分钟之内,而公共绿地、观江阳台、屋顶平台等的步行可达时间大部分在 10 分钟内(图 17)。

(2) 生态肌理完整度:通过 ArcGIS 的生态肌理分析,发现生态基质密度为 30.4%,连接度为 0.43,呈现出较高的肌理一体化特征[23];同时可以看到,在生态绿脉限定下各住区占地面积

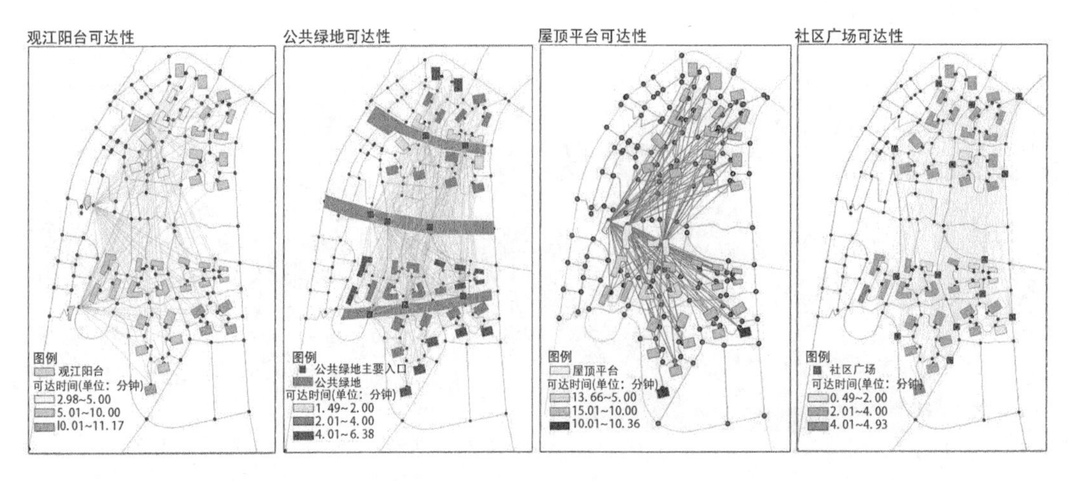

图 17　城市设计方案的公共空间可达性分析
(来源：作者自绘)

适中(最大 3.31 公顷)，具备较高的开发弹性，符合“精明增长”导向及建设“街坊式住区”的要求[31](图 18)。

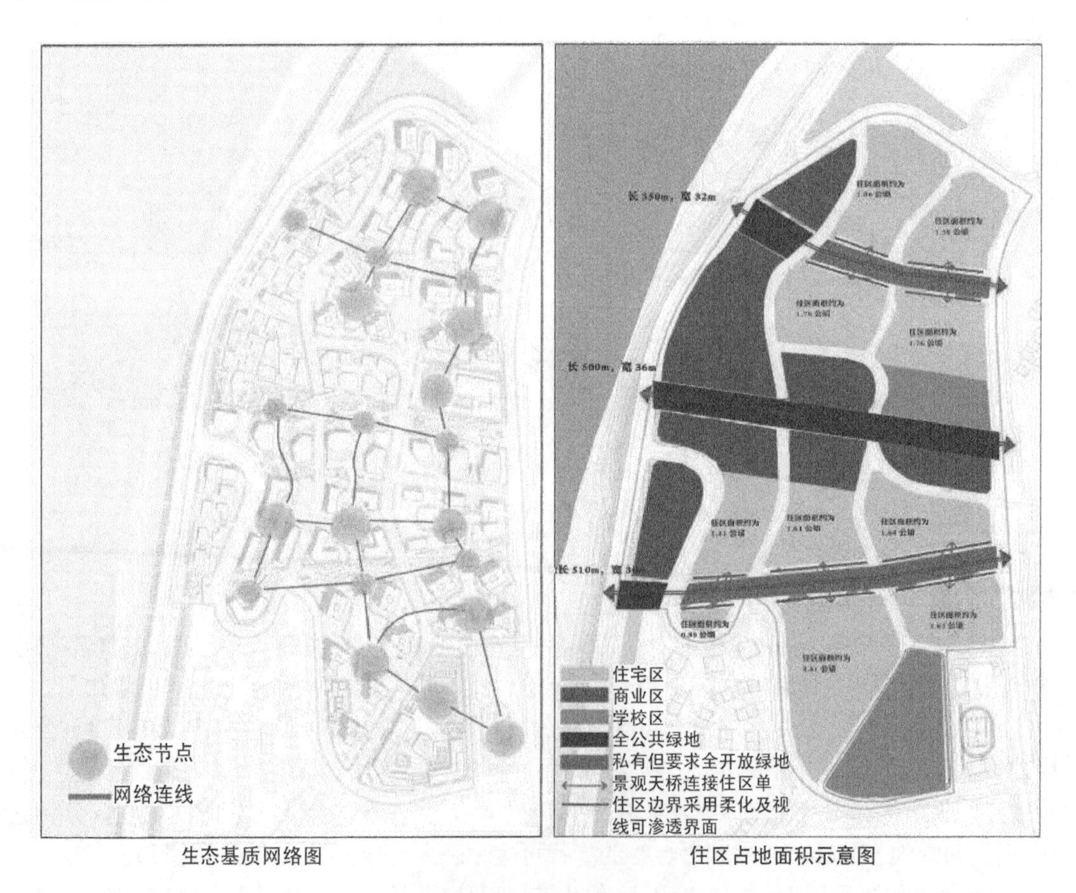

图 18　城市设计方案的生态肌理分析(左)及绿脉限定住区占地面积示意(右)
(来源：作者自绘)

(3) 空间形态适应性:笔者将其分为社会适应性、交通适应性、形态适应性 3 个指标,其中:①社会适应性——最主要的公共空间系统"叠翠绿脉"兼容生态维育及社会休闲功能,分布均匀,服务半径覆盖所有社区,避免了生态绿地沦为"点缀空间"或者"边缘空间";②交通适应性——利用 Depthmap 软件对路网结构进行分析,发现交通整合度和人流集聚度最高的地段集中于"多维江岸""叠翠绿脉""峡谷金街"等位置,表明空间形态特征与主要交通需求匹配较好(图 19);③开发适应性——通过 ArcGIS 的土地价值距离衰减分析,发现"叠翠绿脉""多维江岸""峡谷金街""立体公园"等要素构成明显的"开发触媒",使土地溢价提升显著,尤其是临近"叠翠绿脉"的用地,呈现相较于原控规较大的梯级对比(图 20)。

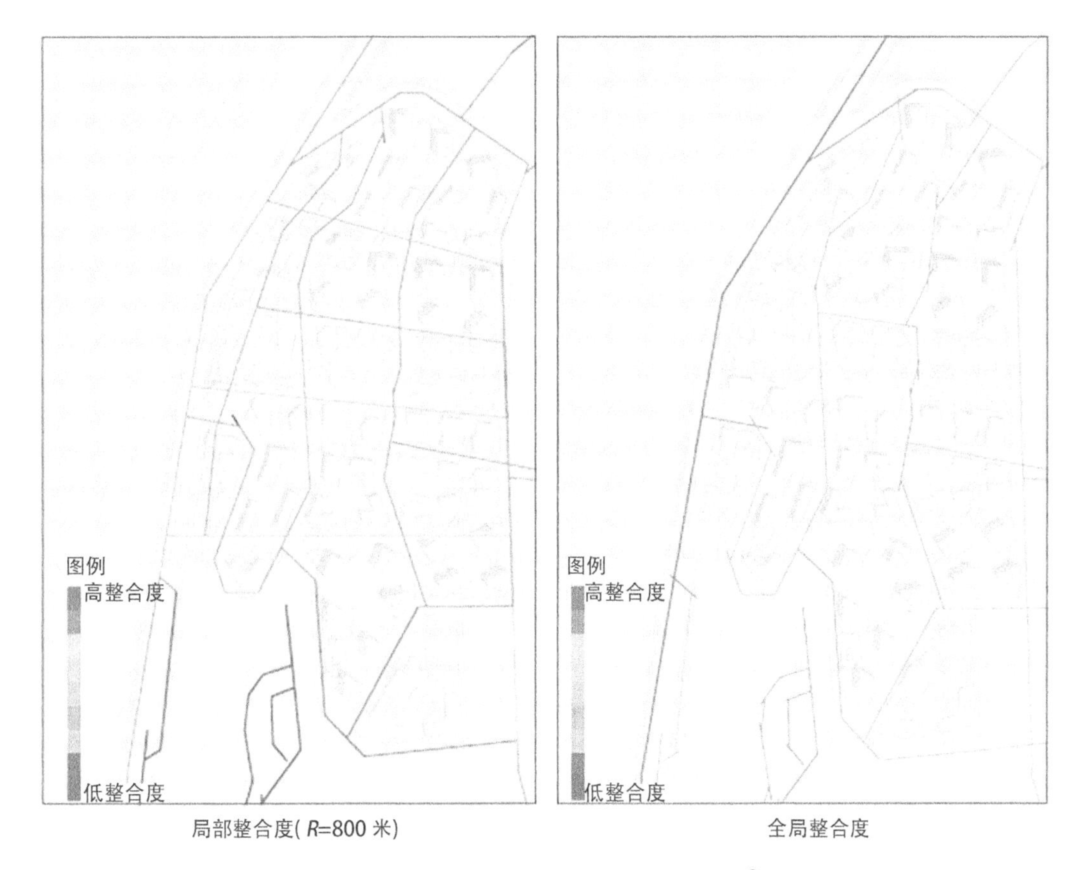

图 19　城市设计方案的交通整合度分析①

(来源:作者自绘)

总体来看,城市设计方案中的公共空间与生态肌理形成较好的耦合,兼具服务社区与促进开发的双重功效;尤其"叠翠绿脉"作为最核心的控制要素之一,体现了"设计要件"及"绿色基础设施"的作用,使整体空间风貌体现出较鲜明的"统领感"及"一体化"特征。

① 局部整合度越高,表明该轴线在周边 800 米步行范围内到达其他点所需的"步数"越少,越易到达;全局整合度越高,表明该轴线在全局步行范围内到达其他点所需的"步数"越少,越易到达。

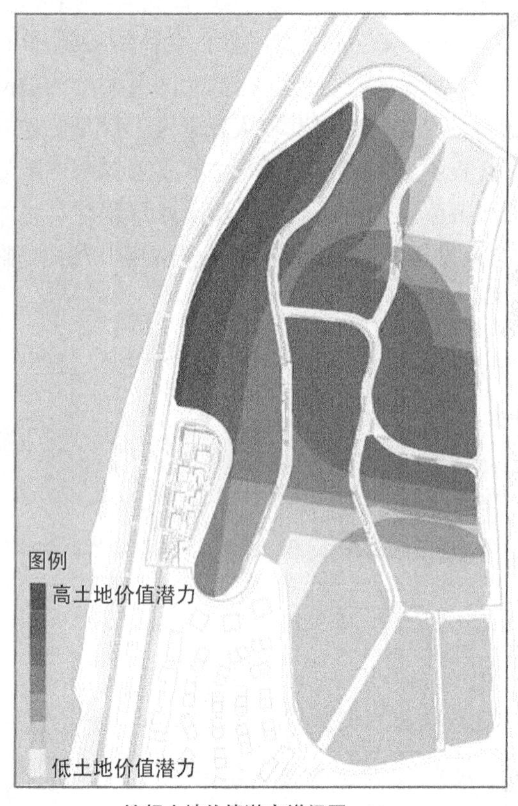

控规土地价值潜力梯级图

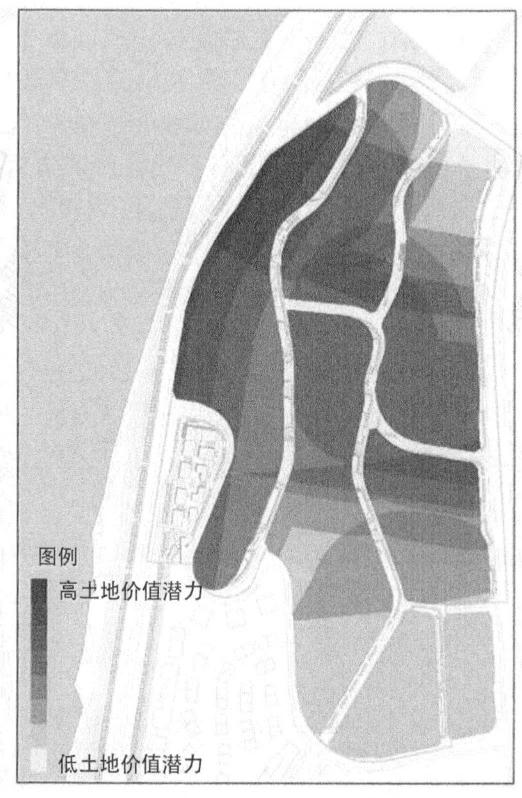

城市设计土地价值潜力梯级图

图 20 控规与城市设计的土地价值梯级对比
(来源:作者自绘)

4.6 城市设计实施路径

一方面,通过“问题—策略—要素”的设计逻辑,本次城市设计构建起一套完整的、能较好体现重庆山地城市风貌和山水自然形胜的城市设计体系,具备向类似尺度和地段的推广价值(图 21)。另一方面,城市设计遵循控规的核心指标体系,主要通过用地布局的重构、整合,及建筑组群和公共空间的分层分段细化控制来塑造整体风貌,形成“控规控指标、城市设计控形态”的“双控”方式,由此明确了城市设计的管控边界和重点,实现了与上位规划的合理衔接。

在此基础上,城市设计的实施路径十分清晰:

(1) 步骤 1:调整控规的土地利用图。由于不涉及核心开发指标的改动,这种调整在现行规划管理体制内被视为控规的“一般技术性内容修改”,在程序方面较为简单易行,避免了烦琐的技术论证及相关部门的利益博弈,有利于提高行政审批效率(图 22)①。

① 据本项目及其他一些城市设计的经验,重庆市政府发布了《重庆市人民政府关于进一步加强主城区控制性详细规划修改管理工作的通知》(渝府发〔2016〕14 号),明确了控规“一般技术性内容修改”与“强制性内容修改”的各自前提及程序差异,这是重庆市城市设计试点工作取得的重要成果之一。

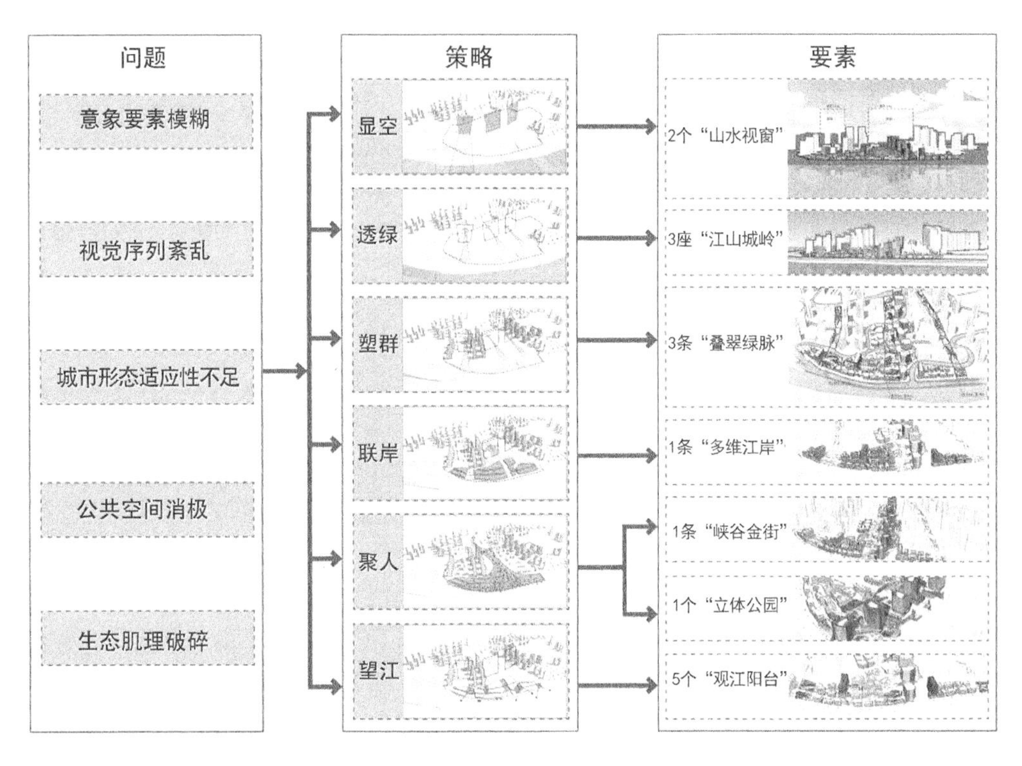

图 21　以“问题—策略—要素”为构成逻辑的城市设计体系,具备适应本地地域条件的系统性和普适性

(来源:作者自绘)

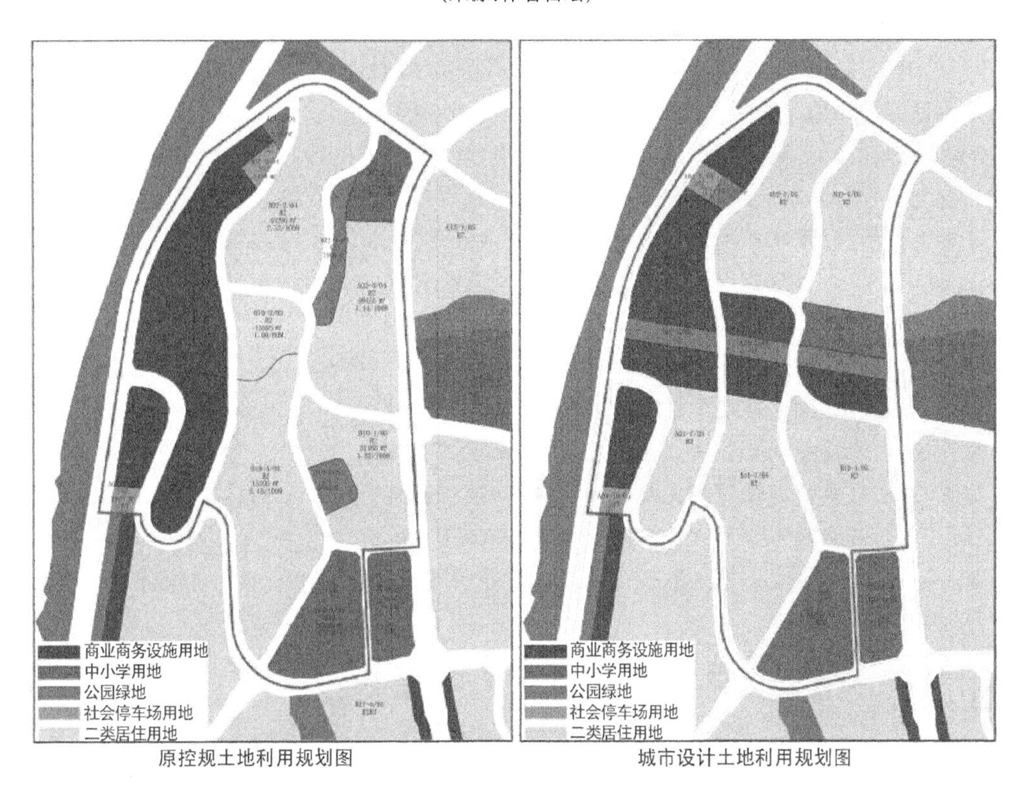

图 22　土地利用图调整前后对比(左:控规;右:城市设计)

(来源:作者自绘)

(2) 步骤2:制作城市设计控制图则。以“控制要素引领”为纲,将城市设计最主要的控制内容转化为简明清晰的控制图则(即将“设计语言”转化为“管理语言”),图则与控规框架下的规划用地许可函合并成一套完整的文件,作为土地出让时的法定要求。

(3) 步骤3:纳入建筑设计方案审查。在土地出让后的建筑设计方案审查阶段,核心开发指标体系参照控规要求,风貌与形态方面参照城市设计要求(遵循“双控”原则)。如需修改,则依据法定程序重新进行城市设计论证。

5 总结

在后城市化阶段,中国城市面临普遍的城市风貌危机,而城市设计被认为是化解风貌危机的主要手段之一。本文提出“意象系统”“视觉序列”“形态适应性”“公共空间”“生态肌理”是构建城市风貌的5个核心维度,城市设计的风貌塑造应该基于这5个维度的正向凸显及一体化整合。

山地城市在地性特征明显,但体现出“整体意象紊乱、视觉认知失序、城市形态与自然环境适应性耦合不足、生态肌理碎片化及生态空间社会功能失落”等问题。本文以重庆为案例,阐述了“以风貌问题为导向——以策略推导总体方案——以要素归纳导控要求”的城市设计实践过程。根据对5个维度的模拟测度,显示对城市风貌的塑造达到较好的实效。

本次实践研究的意义还在于探索构建起一套适应本地地域特征、具有一定普适性的城市设计管控体系和路径。它遵从真实的城市管理机制和开发逻辑,符合当前城市治理的实际需要,可推广应用于类似环境下类似尺度的地段,体现了“做有用的城市设计”的价值导向①。当然,在当前国家整体规划体系向国土空间规划转型的大背景下,城市设计的操作方式及其与既有法定规划的关系或也不可避免面临新的变革和重构;但是在全域空间规划的视角下,对不同尺度的城市风貌(乃至更广泛的建成环境与非建成环境风貌)的管控、提升、修复仍将是一项长期的目标与工作,而城市设计在其间的作为与价值,需要持续的研究与探索。

本文案例项目《重庆巴南区花溪街道滨江重要地段城市设计》获2017年度全国优秀城乡规划设计二等奖(中国城市规划协会)、2016年度重庆市优秀城乡规划设计一等奖(重庆市城市规划协会)。

作者单位:

杨震,重庆大学建筑城规学院教授、博士生导师。

朱丹妮,广东省建筑设计研究院有限公司,助理工程师。

陈瑞,中国城市规划设计研究院西部分院,助理工程师。

汪乐,重庆筑恒城市规划设计有限公司,工程师。

项目完成单位:重庆大学规划设计研究院有限公司、重庆筑恒城市规划设计有限公司、重庆市规划设计研究院。

该项目得到张睿、李峰、余露、刘睿、谢力、蒋笛、王伟涛、陈春、黎柏麟、洪霞等同志的支持,谨致谢意。

---

① 呼应了2017年住建部城市设计试点工作座谈会提出的要求:新时期城市设计要以“问题和目标为导向”,要“解决城市存在的具体问题”。

图片资料均由作者拍摄或绘制,或从项目成果中提取制作。

参考文献

[1] 杨华文,蔡晓丰.城市风貌的系统构成与规划内容[J].城市规划学刊,2006(2):59-62.

[2] 王建国.城市风貌特色的维护、弘扬、完善和塑造[J].规划师,2007, 23(8):5-9.

[3] 杨昌新,龙彬.城市风貌研究的历史进程概述[J].城市发展研究,2013, 20(9):15-20.

[4] Taylor N. The elements of townscape and the art of urban design[J]. Journal of Urban Design, 1999, 4(2):195-209.

[5] Jivén G, Larkham P J. Sense of place, authenticity and character: A commentary[J].Journal of Urban Design, 2003, 8(1):67-81.

[6] 赵燕菁.城市风貌的制度基因[J].时代建筑,2011(3):10-13.

[7] Sepe M, Pitt M. The characters of place in urban design[J]. Urban Design International, 2014, 19(3): 215-227.

[8] 杨震,刘欢欢. 当代中国城市建筑的"迪斯尼化":特征与批判[J].建筑师,2015(5):69-74.

[9] 杨震. 消费时代的建筑美学[J].建筑师,2017(4):105-110.

[10] Barnett J. An introduction to urban design[M]. New York: Harper and Row, 1982.

[11] Punter J. Developing urban design as public policy: Best practice principles for design review and development management[J]. Journal of Urban Design, 2007, 12(2): 167-202.

[12] Carmona M. The place-shaping continuum: A theory of urban design process[J].Journal of Urban Design, 2014(1):2-36.

[13] Lynch K. The image of the city[M]. Cambridge, MA: MIT Press, 1960.

[14] Cullen G. The concise townscape[M]. New York: Van Nostrand Reinhold Company, 1961.

[15] Lynch, K. A theory of good city form[M]. Cambridge, MA: MIT Press,1981.

[16] Alexander C, Ishikawa S, Silverstein M. A pattern language: Towns, buildings, construction[M]. New York: Oxford University Press, 1977.

[17] Rowe C, Koetter F. Collage city[M]. Cambridge, MA: MIT Press, 1978.

[18] Krier R. Urban space[M]. London: Academy Editions, 1979.

[19] Relph E. Place and placelessness[M]. London: Pion Limited, 1976.

[20] Norberg-Schulz C. The phenomenon of place [J]. Architectural Association Quarterly, 1976(4): 310.

[21] Oldenburg, R. The great good place[M]. New York: Marlowe & Company,1999.

[22] Mostafavi M, Doherty G. Ecological urbanism[M]. Cambridge, MA: Harvard University Graduate School of Design, 2010.

[23] Waldheim C. Landscape as urbanism: A general theory[M]. Princeton: Princeton University Press,2016.

[24] Barton H, Grant M, Guise R. Shaping neighbourhoods: A guide for health, sustainability and vitality[M]. London: Spon Press,2003.

[25] Farr D. Sustainable urbanism: Urban design with nature[M]. Hoboken: John Wiley & Sons, Inc., 2008.

[26] 杨震,费保海,郑松伟,等.基于存量环境更新的山水城市总体城市设计:开县案例[J].城市规划,2016

(3):51-57.

[27] 杨震.范式・困境・方向:迈向新常态的城市设计[J].建筑学报,2016(2):101-106.

[28] Carmona M. The formal and informal tools of design governance[J]. Journal of Urban Design, 2016, 122(1): 1-36.

[29] Carmona M. Explorations in urban design: An urban design research primer[M]. Farnham: Ashgate, 2014.

[30] 黄海静,陈纲.基于分形理论的城市滨江地带风貌控制：以重庆市两江四岸地带为例[J].新建筑,2015(3):120-124.

[31] Adams D, Tiesdell S, White J T. Smart parcelization and place diversity: Reconciling real estate and urban design priorities[J]. Journal of Urban Design, 2013, 18(4): 459-477.

“共享生活”理念在大都市城市更新中的探索

The exploration of the concept of “shared life” in the renewal of metropolitan cities

张亚津
Zhang Yajin

1 中国大都市的“共享生活”可能性

1.1 历史城区城市更新的巨大挑战

中国大城市老城区普遍正在面临城市更新的挑战。虽然交通与服务设施优良,但传统街区中人居环境条件差,社会生态退化,是历史城区面临的典型问题。

对比其他城市,北京旧城改造面临更为严峻的社区下滑问题。基于街区诊断的大栅栏地区城市更新路径研究,徐勤政等指出:

- 5.62万户籍人口,实际常住人口为3.61万,以及1.27万的巨大流动人口总量。
- 整体区域户均住房21平方米,人均住房面积不足10平方米。
- 大栅栏街区78%的家庭没有独立卫生间,26%的家庭没有厨房(含自建)。

报告由此指出:大栅栏街区正在转化为高密度、老龄化、低收入、弱势群体集聚的城市底层蜗居地。

传统北京四合院以院落为单元的居住结构,早已转化为以“间”为单元、平均面积为12平方米的户型。对于一个家庭而言,这一空间仅能容下卧室而已,大量传统建筑无法容纳附属厨卫。大杂院中的加建建筑,以厨房、卫生间、浴室为主,事实上是被迫的共生服务空间。

户均和人均建筑面积过低,同时造成自我更新缺乏弹性空间,超越一般居民自身能力。加上院落中环境恶化,街区层面缺少民生基础设施。老城区房租水平因此大幅度低于城市近郊地区。此外,作为历史街区,开发利用方式受限,缺乏整合利用的思路。

这一情况并非仅在北京出现。上海石库门里弄的人均住房面积仅为5.8平方米。进行有机更新、提升居住条件是本地居民的需求,也是历史城区发展文旅产业、服务产业、商贸经济的迫切需求。

对应上述背景,北京旧城改造的现有情况是:

✓目前北京旧城区各个历史街区，普遍采取腾退或平移的方式，腾挪出整院作为更新单元，整体作为商业单元出租或出售，它对以“间”为单位的腾退，提出更高要求；而迁入的办公机构/商业结构，形成社群的异体结构，对原有的社群并无太大益处。

✓以拆迁型费用标准——以每平方米10万以上价格来腾退直属公房，代价巨大，难以持续性应用，未来北京申请式退租或许面临相似的问题。反之，这造成居民没有动力进行自我升级改造，只会消极等待，包括私房在内的整体社区环境进一步恶化。

✓在北京整体旧城策略背景下，原有社群已大批迁出，现有社会阶层正在下滑，新人群与传统社群关联度很低，拜访型人群干扰居住职能。

✓整体社群迁出后，作为商业项目进行整体改造的道路越来越受到质疑，鲜鱼口、大栅栏等商业项目并未取得整体性社会认知。

在徐勤政、何永等学者进行的大栅栏街区调研中，36%的家庭明确表示不满意现状，74%的居民承认难以自主迁房，92%的居民都表示愿意接受腾退，但5.62万户籍人口中，2011—2015年间实际签约的仅为614户。旧城改造的巨大代价，使各种传统型的整体改造主体都对其望而却步。

1.2 青年群体：大都市活力的主体人群

在另一个视野，大城市的青年群体，正在成为居住水平高度挤仄，同时通勤压力巨大，需要临近中心城市居住可能性的另一群体。

租房、合租是青年一代居住的重要选择。

目前，中国大都市的住房市场已经开始从“买不起”向“租不起”转变。以北京为例，2018年7月，北京房租同比上涨达21.89%，以每平方米92.33元的价格居全国之首。北京租房人数占比达37%，约有800万，按照一个人租房的开销占收入比例计，北京的“房格尔系数”约为58%，约为居民可承受的房租收入比(25%～30%)范围的两倍，房租压力过大。

根据2018年北京中心城区人均房租与薪资占比情况，主要就业地所在的中心城区均超过了50%，东西城区两区房租最高，而收入水平最低，租金占薪资比达到65.4%和76.4%。这一数据在国际各大都市都处于高位(图1、图2)。

根据国家统计局和中国房产信息集团(CRIC)的数据：北京、上海或深圳一套普通100平方米住房的按揭贷款需要千禧一代工薪阶层平均收入的1.8倍来偿还。同等价值房屋的租金约为租房者收入的58%，如果是合租的话则为租房者收入的25%。

与此同时，理想的居住环境是青年一代的重要价值取向，中国社会科学院的一项研究对1 400名应届大学毕业生进行了调查，发现55%的人宁可通过长期租房来维持理想的居住标准，也不愿背上按揭还贷的负担。事实上仅有33%的受访者愿意为了购房而放弃更好的生活方式。2019年的报告显示：超七成青年租客认为不一定要买房，在租来的房子里依旧可以拥有好生活。

最后，中国人更换住所的频率比以往任何时候都高。我国国家卫生和计划生育委员会研究的我国流动人口流动特征数据(2009—2015年)显示，60%的流动人口倾向于长期居住在城市地区，他们在某一特定大城市的预期居住时长为5年。

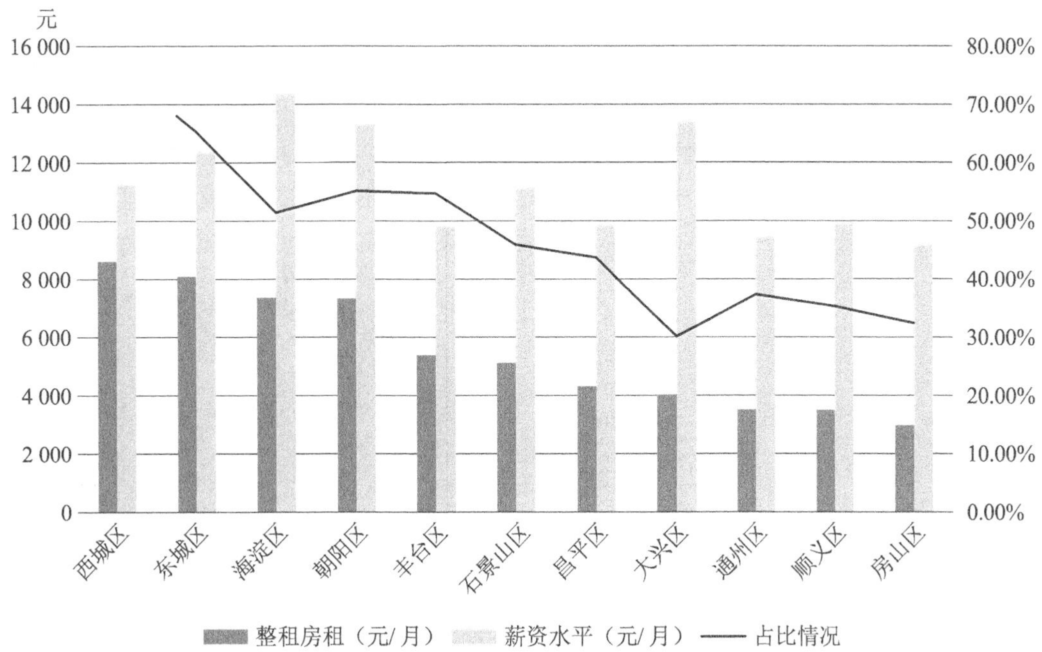

图 1　北京市各区薪资及租金情况

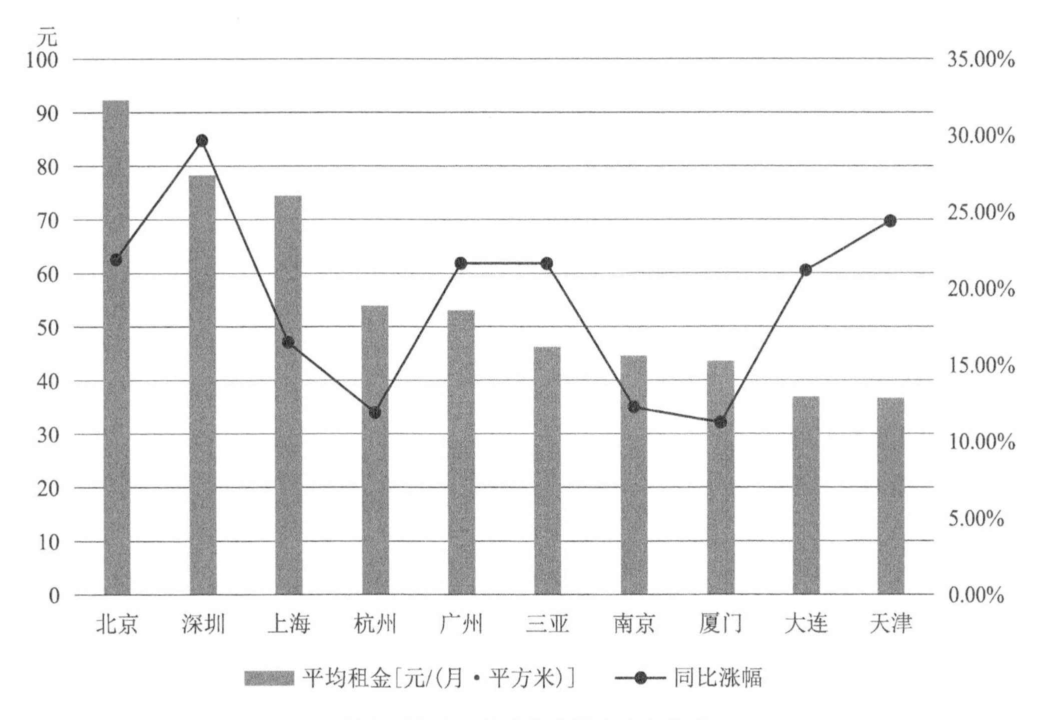

图 2　2018 年全国部分城市房租情况
（数据来源：中国房地产业协会）

大都市旧城区的区位、服务设施与历史文化资源，与青年人群的就业取向、文化生活取向构成了良好的耦合关系。问题在于：如何借助新社群、新产业，与老城区的社会服务共生，与历史文化街区的文化价值共生？

1.3 共享经济的兴起

在共享经济兴起的今天，我们已经成功地共享了工作空间、交通空间，还有什么可以拿来共享？Airbnb之所以可以成为酒店的替代品，除了由于其价格相对更有弹性外，也因为它借助以家庭公共设施、公共空间为核心的共享，赋予了租住者在陌生城市享受家庭生活的便利、租住者与主人之间复合利用的综合价值，以及主人与租住者之间文化交流的可能性(图3)。

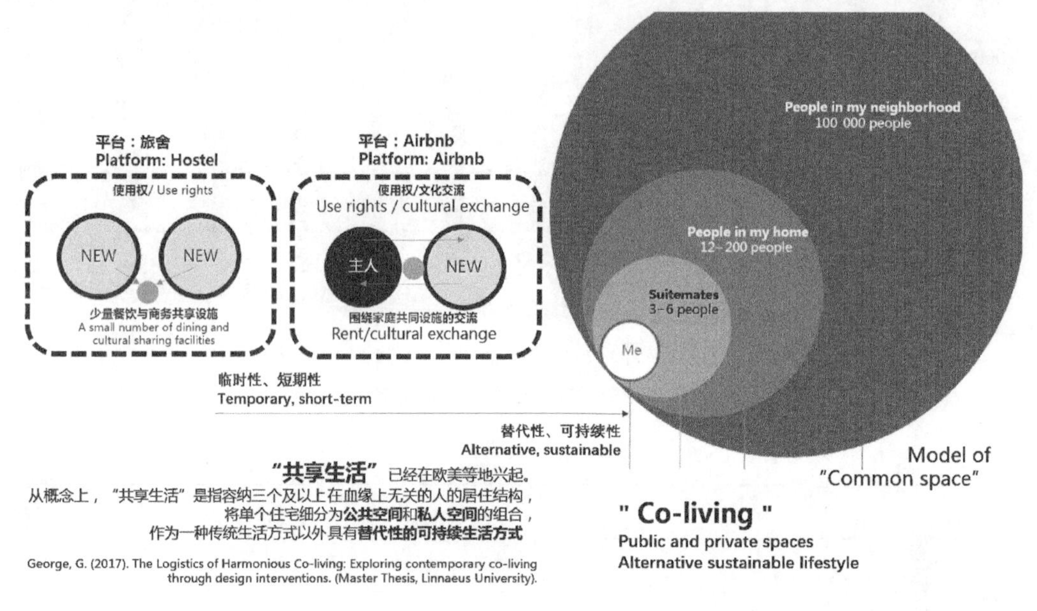

图3 Airbnb与酒店共享空间模式的对比(自绘)，以及新一代共享居住空间的典型模型
(图片来源：www.commonspace.com)

Airbnb仅仅是一个临时性、短期性的解决方案。但是由于上述品质，大多数发达国家在住宅紧缺问题基本解决之后，今天却在研究将Co-living(共享生活)转化为一个替代性(alternative)、可持续性(sustainable)的长期居住模型。这是另一种生活方式，共享生活也许将成为未来的重要可能。George Green在2017年的一项研究界定了"共享生活"是指容纳三个及以上在血缘上无关的人的居住结构，将单个住宅细分为公共空间和私人空间的组合，作为一种传统生活方式以外具有替代性的可持续生活方式。

2017年，中国一些地产集团提出利用城市旧区中的综合资源，为城市中坚力量服务。将长租公寓、联合办公、配套商业有机地结合在一起，构建小综合体生态模式——CityHub，驱动城市中生活服务、文化娱乐、商业配套的资源互通。2019年4月3日，Airbnb宣布，计划在广州市越秀区增加中高端特色住宿消费供给，集中打造新河浦片区"老东山"特色及华侨新村民宿品牌。

被《纽约时报》称为"北京布鲁克林"的鼓楼地区，一直以来是北京国际文化社群的聚集地，沿街随处可见面包坊、早餐店、小画廊，以及周日坐在咖啡馆台阶上聊天的青年人群。大量国际背景的青年社群的自发融入，为一个相对中低收入的历史文化街区的商业业态注入了新的活力，迥异于由中央别墅区、国际五百强总部形成的其他北京国际社区。在这里，历史文化资源与区位构成了核心的吸引力。这是脱离地产机构层面运作,更具自发性的优质新混合型生活空间。

"共享生活"的特殊性意义,或成为未来中国大都市青年新型居住模式的重要可能。

2 关于共享的一系列社会理论模型研究

2.1 共享社会理论的历史回溯与现实意义

分享生活、居住空间的理念在 20 世纪 60 年代曾经一度是"住宅紧缺 + 左翼社会自由主义的理想"融合下的共同产物,一个经过规划的、具有高度社会凝聚力和团队精神的住宅社区被称为有意社区(intentional communities)。我国 50 至 70 年代之后大规模修建的集体公寓,也有类似的空间模型。

1989 年到 1990 年,心理学家丹尼尔·格林伯格(Daniel Greenberg)研究了美国 200 多个有意社区的儿童,他发现在有意社区成长的儿童比在非社区的儿童拥有更多的成人榜样,并倾向于和亲戚以外的成年人建立更多的友谊。社区给孩子们提供了非正式学习的机会,由于接触成人生活,他们往往更"社会成熟、自信、外向、能干和有口才"。社区是成长的好地方,父母可以获得儿童保育方面的帮助和指导,孩子也可以获得玩伴、有趣的设施和父母以外的榜样。

今天的"共享生活"模式,受到主动和被动两方面因素的推动。一方面,社会文化的开放度,促使人们主动地寻求共享思想、共享知识。另一方面,现代社会发展中产生的问题也提出协作共享生活方式的需求。人口增长、生活成本上升、快速城市化进程中出现的空巢老人无人照料、幼年子女无人看护等问题,以及人与人之间的关系疏远、社群交流减少的困境,加上以消费为导向的生活方式造成的资源环境问题等都将促使人们更多地共享生活空间和服务。

"Anton & Irene + SPACE10"发起的 ONE SHARED HOUSE 2030 研究项目对 147 个国家的互联网群体进行网络调研,结果显示:"加强社交"(38%)是选择共享生活的首要原因,加上与之类似的"在工作和学校之外拥有一个社群"(19%),两项社交需求共计占总量的 57%;"消减生活成本并获得更大价值"(21%)、"更好的区位与更好的住屋"(9%)、"日常生活的高度便利"(7%)这三项更为经济化的目标约占 37%,是次要原因;其他选择共享生活的因素还包括"周边有人群可以求助"(4%)、"拥有瑜伽/健身设施"(2%)、"拥有公共空间"(1%)等。

"共享生活"已经在欧美等地兴起。从概念上,"共享生活"是指容纳三个及以上在血缘上无关的人的居住结构,将单个住宅细分为公共空间和私人空间的组合,是一种传统生活方式以外的替代的可持续生活方式。位于瑞士苏黎世的 Hunziker Areal 正是由当地住房合作社发起的一场"共享生活"人居实验,得到了欧洲社会的广泛关注,也启示了一个未来发展的新模型。

2.2 "多代际共享生活"(Co-Generation)

在美国,麻省理工学院(Massachusetts Institute of Technology, MIT)建筑系的两位学生在波士顿发起并运营 Nesterly。背景为:到 2035 年,1/3 的美国家庭将由 65 岁以上的人主导。成员年龄超过 80 岁的家庭中有 57%将选择单人居住。65 岁以上的人中有近 90%的人希望尽可能地住在自己的家中。而且随着房价上涨和通货膨胀，房产税和日常开支将给老年人的生活带来挑战。因此，现在及将来，许多家中有额外空间的老年人，需要外部的帮助以及额外的收入。与此相对，在美国等地，经济适用房的危机使低薪工人和年轻人难以找到居住的地方。

Nesterly 建立了“多代际共享生活”平台，为住房负担能力和老龄化这两大挑战提供解决方案，希望通过多代际共享，构建一个更加相互联系和可支付的世界。老人将家中的空余房间租借给年轻人，两者共享空间、共同生活，年轻人通过为老人提供一定的生活服务来换取房租的减免。Nesterly 平台负责审核出租者和入住者的资料，在合适的申请人间寻找配对，帮助他们就居住的收费、共享生活的规则制定等问题达成共识，并对后续的生活进行一定的跟踪回访，必要时进行调解。接受服务的双方需要支付一定的会费以保证平台的持续运营(图 4)。

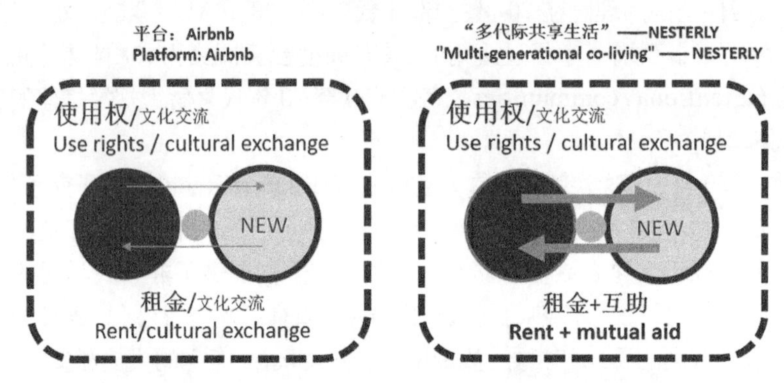

图 4　Airbnb 与 Nesterly 共享空间模式的对比

2.3　“共享生活”长租公寓社区(Common-Shared Life)

在共享经济的整体背景下，全球涌现了大量“共享生活”创业公司，如 Common、Open Door、CommonSpace、Pure House、Founder House、We-Live 等。

Common——共享生活创业公司，在美国纽约、旧金山、芝加哥、华盛顿、西雅图和洛杉矶这 6 个主要城市设有 15 个共享生活居住社区。居住成员涵盖了刚毕业的大学生到已有所成就的专业人士。

社区设计理念：“了解你的邻居”。通过共享的生活空间、公共设施和偶尔的郊游活动提供真正的邻里关系、社区意识。对于租户而言，租金之外，以创新创业为价值观的社群共建与文化交流具有重要的吸引力。

因此，该项目以会员费计费，与传统公寓相比，会员每月可节省超过 500 美金的住房支出。值得指出的是，Common 虽然总体上并不直接盈利，却一直自其五项投资中赚钱。该业务通过收取部分租户的租金和建筑物的净营业收入来创造收入。

观察雪城(Syracuse)一座共享公寓(CommonSpace)的典型平面，可以看到每套 20 平方米的个人公寓呈现居住、厨卫与客厅的小而全布局；个体单元面积较大、功能完整，但空间特色并不突出。

而每层的 30 平方米的共享部分以大型厨房、吧台、游戏和电视空间、洗衣空间为主，并非日常生活情景，功能效率较低。因此，这类长租公寓的居住模式较为单一，人群背景也较为单一,租期通常在 1 年左右。这仍然是一个短期性的、阶段性的小众人群选择。新闻媒体称之为上班族重返"集体宿舍"(图 5)。

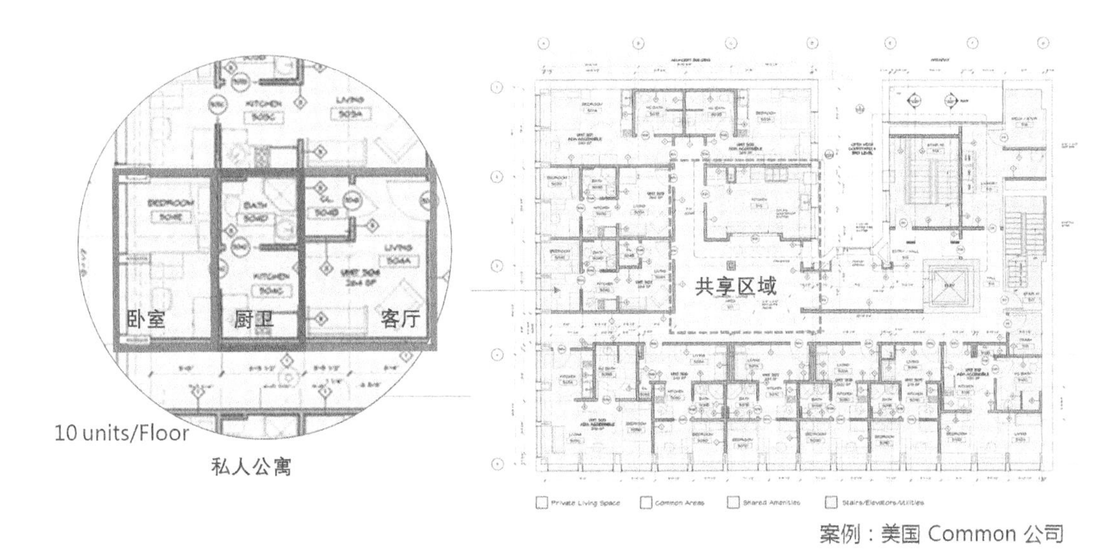

图 5 "共享生活"创业公司格局:小而全的独立公寓

图片来源:CommonSpace 机构网站

2.4 Hunziker Areal——基于"共享生活"理念的居住实验

更具有革新意义的居住模型,已经在瑞士出现。位于苏黎世的 Hunziker Areal 正是由当地住房合作社发起的一场"共享生活"的人居实验,得到了欧洲社会的广泛关注,也启示了一个未来更具有持续性的新人居模型。

Hunziker Areal 的主题是居住之外——为人而建,与人共建(Mehr Als Wohnen, Bauen für und mit Menschen),项目获得了 2011—2015 年苏黎世市优秀建筑奖、2016—2017 年世界人居奖等系列奖项,它在建筑、能源、公共空间、社会等各个领域进行了实验。

Hunziker Areal 位于苏黎世北部边缘,原 Hunziker 混凝土工厂 4.1 公顷的区域内,是苏黎世住房合作社结合 2007 年苏黎世"超越生活/居住之上——百年非营利性住房建设"纪念活动的一个创新性人居实验。2015 年以来,已为 1 200 人提供住房,并提供约 150 个工作岗位(图 6)。

2.4.1 建筑外部:底层公共空间体系与商业公服的全面开放

Hunziker Areal 由 13 栋建筑构成综合社区。

• 主要基地空间公共化,每栋建筑功能复合,底层 6 000 平方米的商业面积中,包括 800 平方米公共建筑面积,承担公共聚会空间、酒店、工作室、办公室、修理车间、桑拿系列职能。地下车库中还有公用冷冻库。

• 独立中庭建筑取代了板楼。建筑间距非常小,最小只有 8 米,大多数在 9 米左右。融合无车化街区,形成了舒适的街道活动氛围。

• 首层商业与公共服务职能向公共广场全部开放,形成了连续、活跃的生活界面;面对绿地的则为居住区域,通过抬高的阳台与公共生活分离。

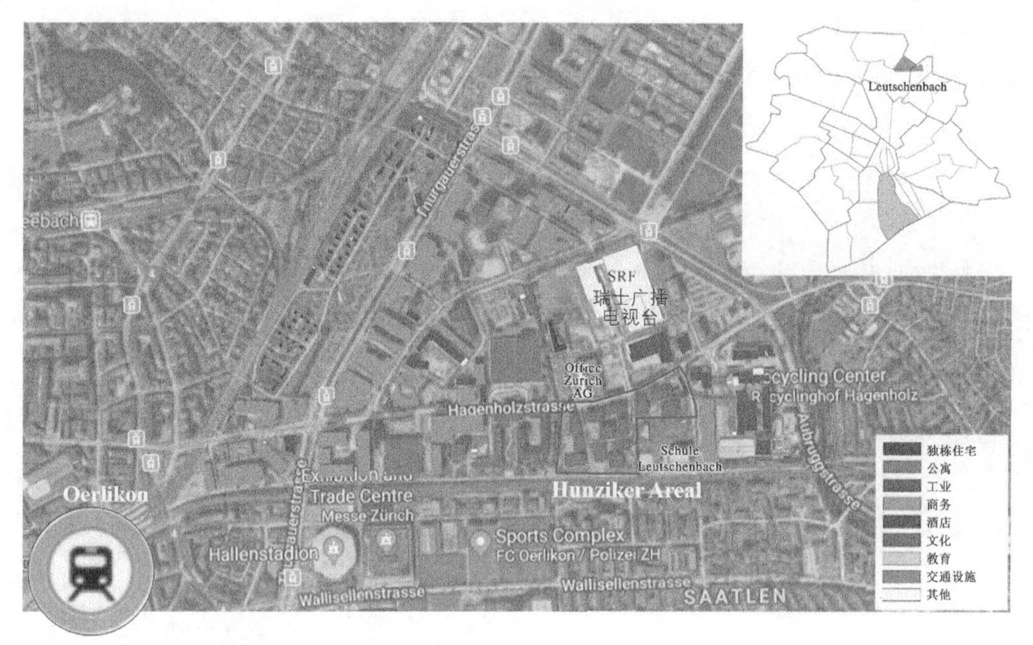

图 6　Hunziker Areal 周边用地

（图片来源：根据 Stadt Zürich 用地情况自绘）

主广场仅仅 38 米×25 米，与 3 个街坊内部的广场构成了具体而微的公共空间等级，尺度上差异较小。相形之下，一系列边缘性的绿地花园用途更为突出，作为与学校融合的开放绿地，和西侧的紧密结构相比，两侧形成了两种差异化的品质。错综复杂的公共空间、通道与公园网络交织在一起，活跃的邻里空间与疏朗的体育绿地都让人印象深刻(图 7)。

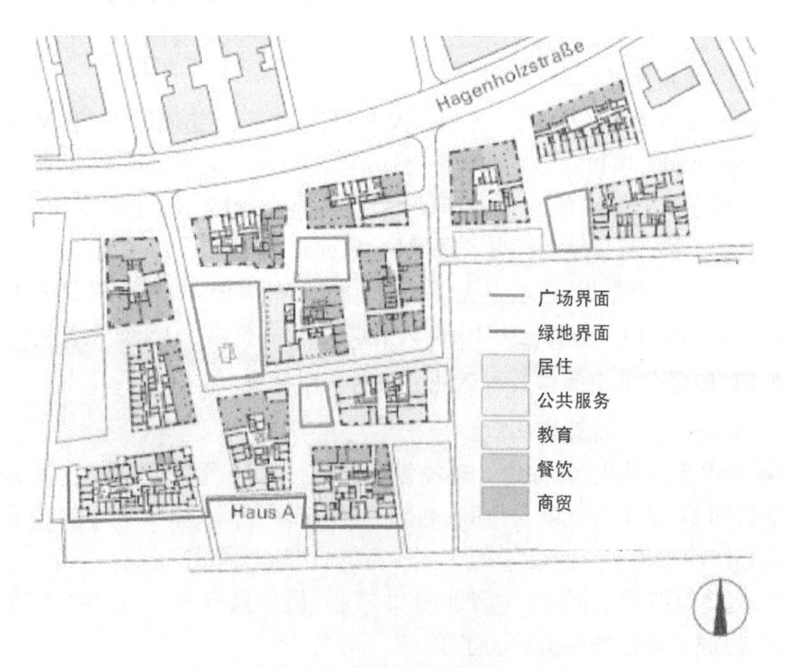

图 7　建筑首层功能以及不同的生活界面

图片来源：DBZ(Deutsghe Bauzeitsghrift)

2.4.2 建筑内部:私人与公共空间的紧密融合

Hunziker Areal 的个体建筑,除私人空间之外,形成了每户、每层、每栋三个层级的共享空间(图 8):

✓部分私人空间仅仅包括卧室与卫生间;也提供包括小型客厅的一居室单元,住户可以选择性地开放共享。

✓每户是一个卫星式套内平面。多个卧室或卧卫单元之间,是连续共享的客厅、浴室、厨房等。公共空间有各种进退形态,形成私密空间和外部公共区域的过渡。各种公寓内部空间的组合形式还可以在建筑物生命周期的更新过程中进行调整。

✓每层设公共洗衣间、晾衣间,缩小套内必需面积。

✓每栋建筑设置中央大厅或宽敞的楼梯间,部分拥有会客厅或公共厨房等公共空间;多数建筑设公共储藏室。由于各户大型公共空间均有大面玻璃采光,在楼梯间就可以感受到各个联合家庭的生活氛围。

图 8　两栋建筑的标准楼层平面图

(图片来源:Duplex Architekten, FUTURAFROSCH GMBH)

Hunziker Areal 项目提供的不是两室一厅、三室一厅的标准住宅,而是从 1 室到 7.5 室,甚至 12.5 或 13.5 室一“厅”的共享住宅。各种公共空间使居民联系紧密,为一群“想要共同生活但又想要退却的人”提供多样化的居住、多样化的社群选择(图 9)。

该项目鼓励建筑师探索新的规划和组织模式,为人们提供各种合作共享的新方式。因此每一栋建筑的形态都是完全不同的,13 栋建筑因此形成了极其多样化的居住可能。部分建筑还特别设计了与公共空间衔接的空间,例如 Dialogweg 6 & 2 的阶梯形公共空间。

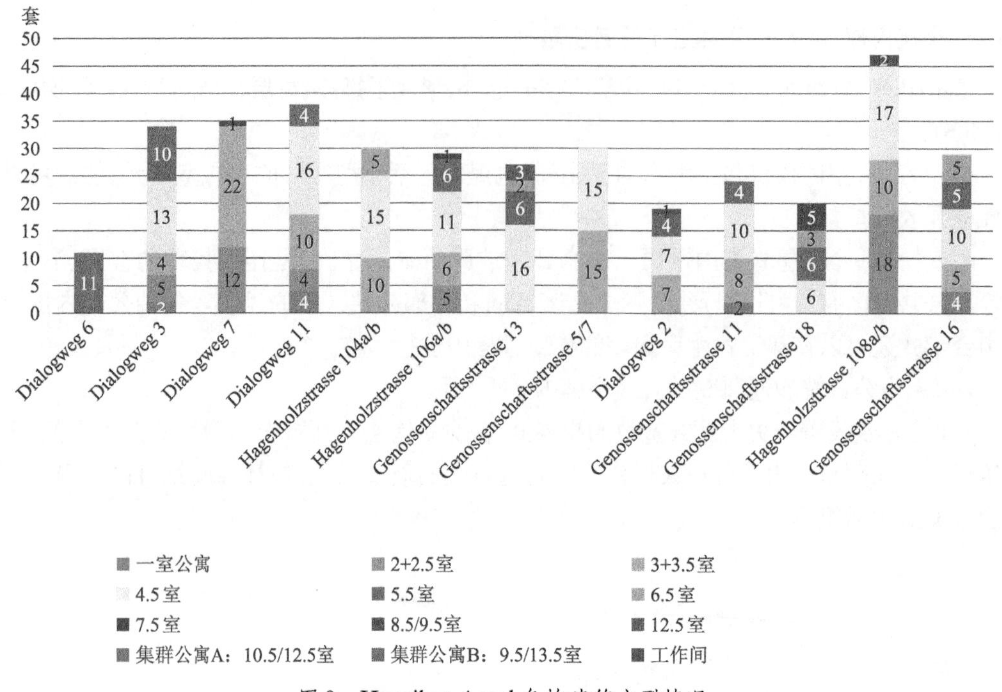

图 9　Hunziker Areal 各栋建筑户型情况

(数据来源：Baugenossenschaft mehr als wohnen)

实体空间以外，可通过 Hunzikernetz 的内部网络平台进行信息、商品和服务的交换、出售和赠送。此外，开放性的二手交易平台也体现了社区广泛合作、居民互相信任的特色。

此外，Hunziker Areal 加入了瑞士提出的 2 000 瓦社会的能源政策愿景，以节约资源的方式建设和运营社区，促进可持续的生活方式，共享生活空间。

Hunziker Areal 将平等、共存、合作、融合、服务等原则纳入租赁守则，鼓励居民在年龄、背景、经济和专业等方面的多样性。从单身青年、夫妻、带小孩的家庭到老年人，从安置的难民到中等收入的专业人士等，都可以在社区共存，享受生活和社区服务。Hunziker Areal 的成员特征中，家庭占比近 50%，青年人与中年人占比约 30%，老年人与大学生总量约 20%。分析苏黎世的平均情况可以看出 Hunziker Areal 对年轻人与家庭的吸引力。34 岁以下的中青年人达到了 68%，这是一个非常年轻的群体(图 10)。

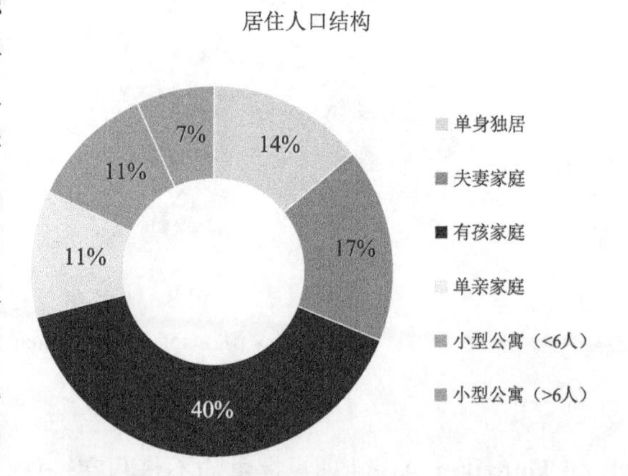

图 10　Hunziker Areal 的居住人口结构

(数据来源：Baugenossenschaft mehr als wohnen)

Hunziker Areal 作为非营利性建设项目，将公寓以实惠的价格(市场价的 70%～80%)租给有资金困难或特殊价值观的家庭。其中 1/5 的公寓得到了低收入补贴。但 20%的补助比例与欧洲大多数新区常常高达 50%的比例相比，事实上较低。

2.4.3 社群:一个新的公共/私人面积比重

Hunrize Areal 私人空间人均面积仅为 15～25 平方米,融合公共职能后的 Hunrize Areal 人均居住面积达到 35 平方米,1/3 的面积以公共共享空间提供。对比苏黎世人均居住面积 50 平方米,这一数字明显更低。与 Common 等长租公寓的模型中共享空间占整体私人空间不足 10%相比,Hunrize Areal 的共享空间占每个单元个人空间的 30%。

共享空间的比重决定了个人生活的比重,Hunrize Areal 因此呈现了与 Common 长租公寓生活方式的巨大不同。Hunrize Areal 外部和内部提供了不同层次的聚会和社交机会,深层的共享空间鼓励近距离的生活和工作方式。私人、半私人和公共空间之间的界限开始模糊和重叠,这为新的生活方式提供了更多的可能(图 11)。

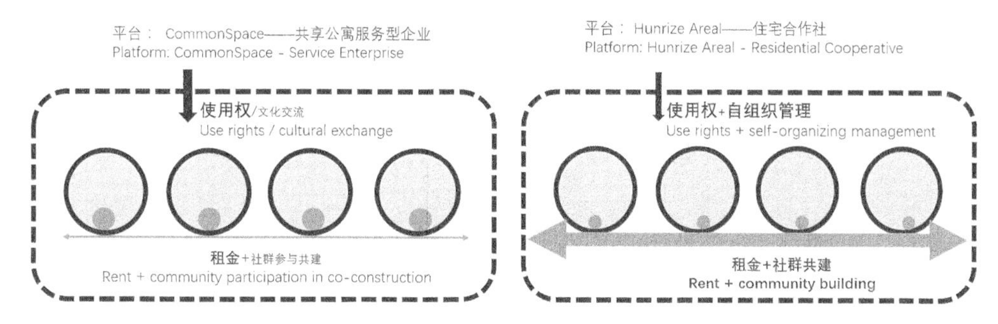

图 11 CommonSpace 与 Hunrize Areal 共享空间模式的对比
(图片来源:自绘)

相似的案例正在西欧各国被尝试,例如慕尼黑共管式住房 WagnisArt。现代的"共享生活"居住实验空间正在得到新一代城市管理者的广泛关注。

3 北京法源寺胡同的"共享街区"更新——共享街区的进一步塑造

法源寺历史街区为北京 33 片历史街区之一,位于北京二环内西南侧,紧邻两处地铁站点,区位条件优越。法源寺自唐代始建,是中国佛学院的所在地。现存建筑为国家级文保单位。街区内共有 24 处会馆遗迹,其中 4 处为市区级文保单位(湖南会馆、浏阳会馆、绍兴会馆、粤东新馆),片区内聚集了以牛街为代表的北京民俗文化、佛教文化、会馆文化、伊斯兰文化等极具特色的传统文化遗产信息。

与此同时,街区居住拥挤,现状常住人口为 4 703 人,人口密度约为 2.9 万人/平方千米,人均居住面积(住宅建筑面积与常住人口)约 10.4 平方米,拥挤型院落、特挤型院落占街区院落总数的 51%,容纳了常住人口的一半。至 2019 年底,西城区平台公司腾退房屋 301 户,释放 427.5 间已腾退空房,小而零散,没有可利用的完整院落。

法源寺历史街区更新中,鲜明提出了"共享街区、共享生活"的思路,在现有局促的街区中,少量迁移,大部更新,融入多级共享公共设施,成为现实而有公益意义的解决方案。借助街区基础单元与公共设施的提升，鼓励青年人、青年家庭，以更加融合性的方式深入历史街区中的传统社会，更加有机地逐步置换新鲜血液，并形成历史街区的坚实人口基层。

整体目标在于：最大限度地降低投资，集中高效改善留下来的原住民的生活条件，同时引入新的产业和活力空间，为老城注入新的发展活力。

法源寺历史街区将着力构建三个层面的价值共享：

第一层级：共享街区

以整个街区作为共享空间，在保护历史风貌的同时，由公司统一进行运营管理、渐进型改造。通过在整个街区高水平地布局社区公共职能，将街区本地的日常生活变成新兴都市服务业区域、青年人群能够融入的都市家园，并融合颇具特色的体验式旅游产品。

东西两区采用不同的改造思路。东侧片区临近地铁出入口和菜市口大街，区位交通优势明显。由于地铁四号线的影响，很多居民有腾退意愿。同时较多的名人故居、会馆是东区深厚的文化基础。故而东侧片区着重组织整院改造，作为办公、酒店、公共文化设施等用途，为街区提供稳定的经济产业来源与潜在的就业空间(图 12)。

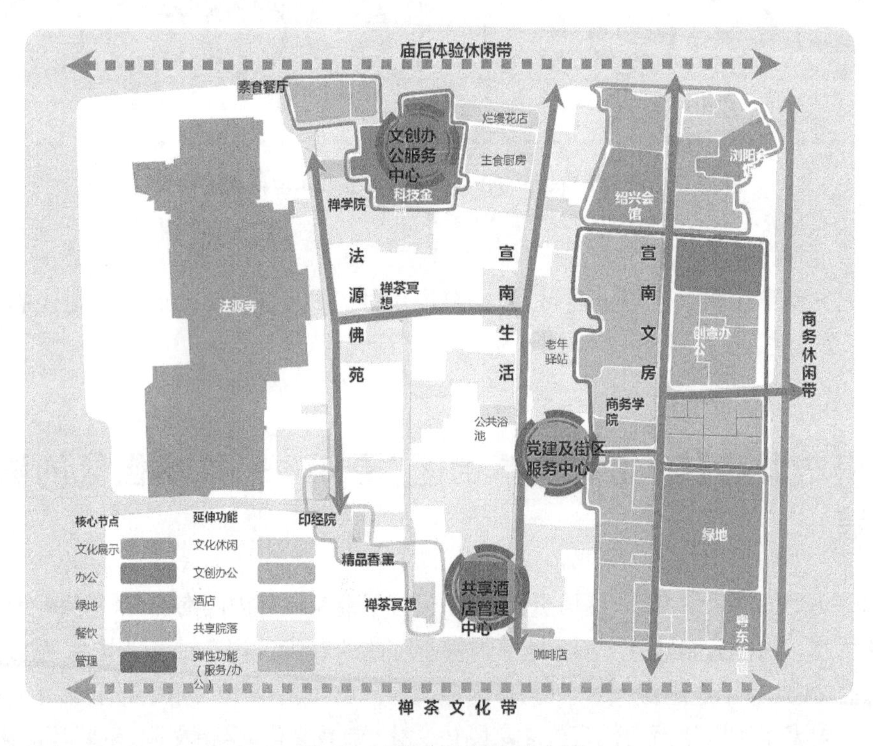

图 12　法源寺街区改造分区

（图片来源：自绘）

西侧片区，即烂缦胡同片区，保留有较为完整的老北京民俗风貌。这一区域将利用灵活的功能规划将零散的空间组织起来，构建高品质文旅目的地，提升居民的生活质量，以烂缦胡同整个片区发展共享街区(图 13)。

• 对 400 多间已腾退房屋，通过插入公共服务和商业性功能单元，探索“平房成套化”的单间更新模式。

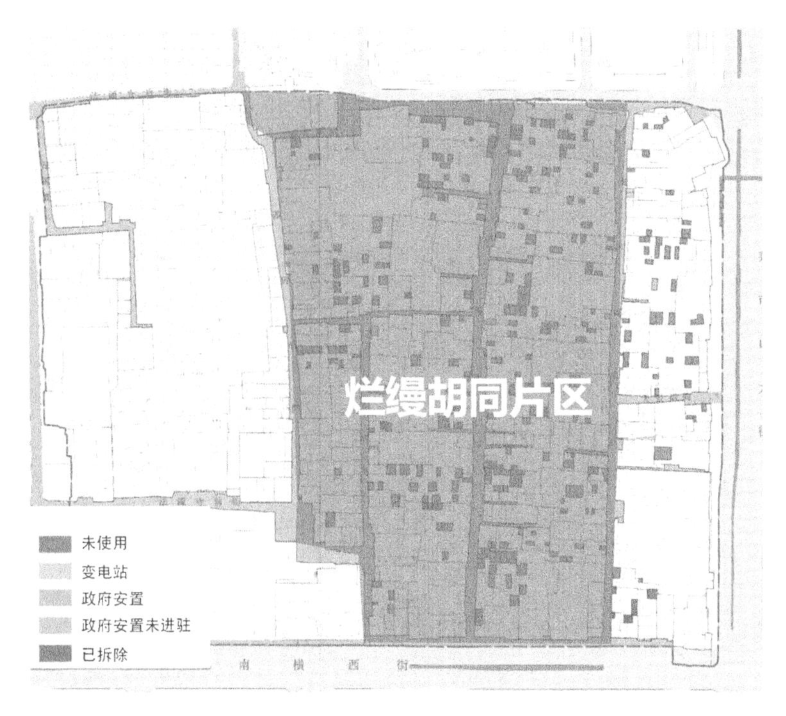

图 13　西区及街区内已腾退房屋分布情况(自绘)

• 由地产公司结合中心酒店,统一作为"胡同中的客房""胡同中的青年公寓"单间管理,整体服务。

• 提供社区中高水平的公共服务设施,同时成为酒店、文旅、青年公寓共享的社会服务与社会交流平台(图 14)。

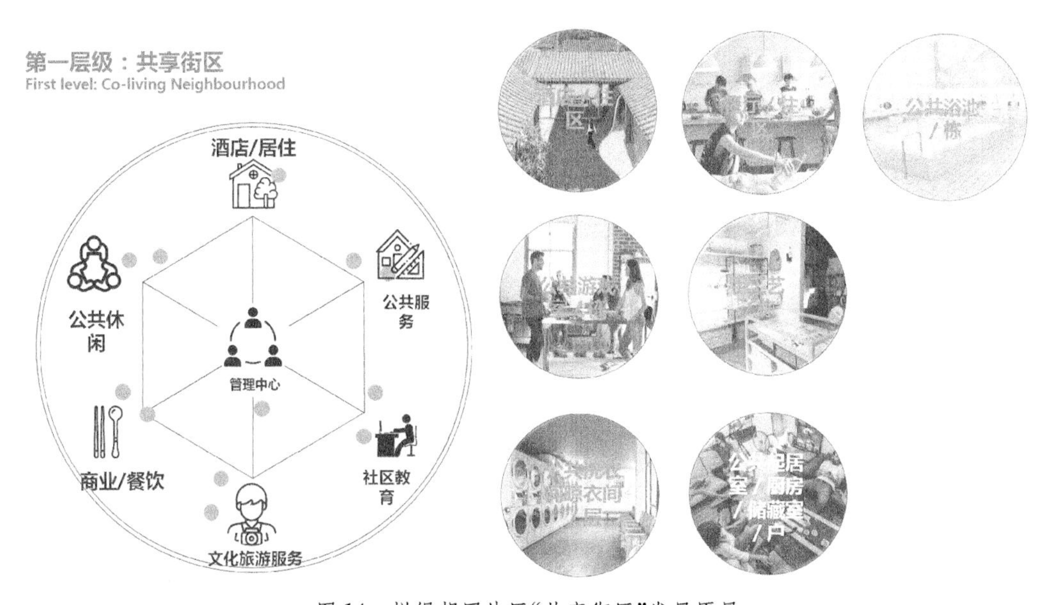

图 14　烂缦胡同片区"共享街区"发展愿景

第二层级：共享院落

以四合院为整体，保护历史风貌为前提，引入长租公寓类型的稳定社群，同时融合公共服务配给；以微改造方式，按照院落内未使用房屋数量和大小，插入不同功能的共享空间，配套相应设施和服务，整体提高社群活力与生活品质(图 15)。

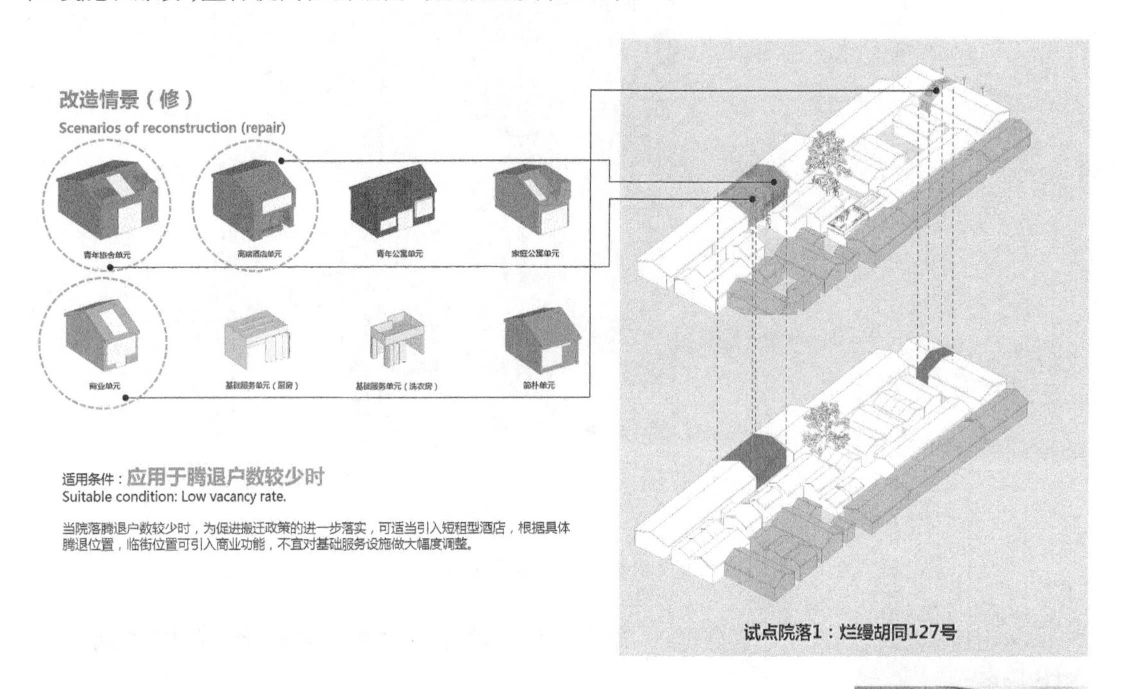

图 15 院落内的共享空间模型与公共服务设施

（图片来源：自绘）

—— 基础配套功能:部分 12～15 平方米左右的房屋,改造为院落居民共享的公共设施,配套卫生间、公共厨房、洗衣房等民生基础设施,提升居住质量。

—— 居住服务功能:引入专业民宿、公寓品牌机构,合作改造经营街区院落内 10～20 平方米的绝大部分未利用房屋,卫生间(选加厨房)入户使住宅成套,以分散型小微住宅为青年群体、渴望体验老北京生活的文旅人群提供居住、酒店服务,构建街区商业经济引擎。

—— 环境整治与共享服务:院落内,拆除临建后,进行公共空间环境的整治与共享服务设施的融入设计。

—— 商业服务功能:部分房屋进行街区内的商业经营,为街区居民和访客提供休闲娱乐、健康养生等服务,构建街区公共服务空间。

第三层级:共享建筑

借助高品质建筑设计的融合组织对建筑内部进行整体改造,在保护历史风貌的前提下,配建基础服务设施(图 16)。

- 对 12 平方米左右的典型单元,利用垂直净空与集合家居空间设计,给予青年家庭、多代居的完整职能。
- 2～3 个 12 平方米左右的单元联合,特别设计为青年公寓。
- 12 平方米左右的临街面房,转化为面包店与咖啡店。控制开放度,以保障传统居住街区的空间特色。
- 少量单元转化为深具北京文化特色的民宿酒店优质单元。

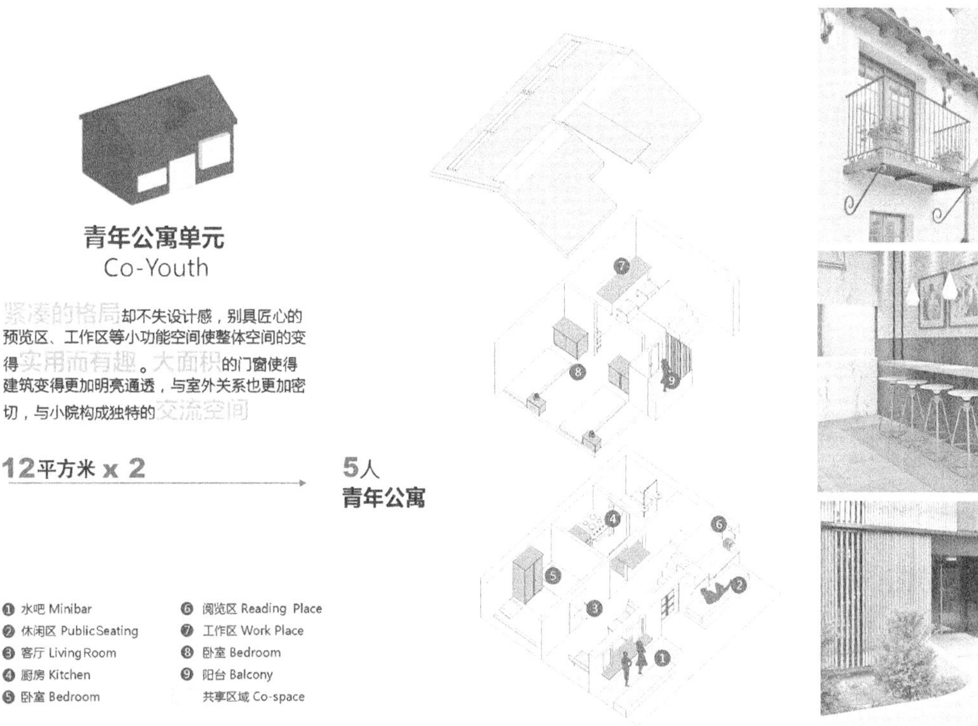

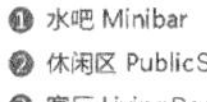

家庭公寓单元
Co-Ly

独立院落带来更高品质的生活环境，利用条件可以创造独特的功能空间。能完全打开的大门可以瞬间将室内外连通，将院落生活引进室内

12平方米 → **2**代**3**人 **年轻家庭**

❶ 灵活开敞空间 Public Space
❷ 入口 Entrance
❸ 厨房 Kitchen
❹ 卧室 Bedroom
❺ 洗手间 Bathroom
❻ 卧室 Parents' Bedroom
❼ 儿童房 Children's Bedroom
❽ 阳台 Balcony
共享区域 Co-space

商业单元
Co-Fe

❶ 公共休闲区 Public Zone
❷ 吧台 Cafe Bar
❸ 底层卡座 Seating
❹ 楼梯(下) Down
❺ 上层卡座 Seating Up
❻ 绿植墙 Green Wall
景观窗 Folding Window
共享区域 Co-space

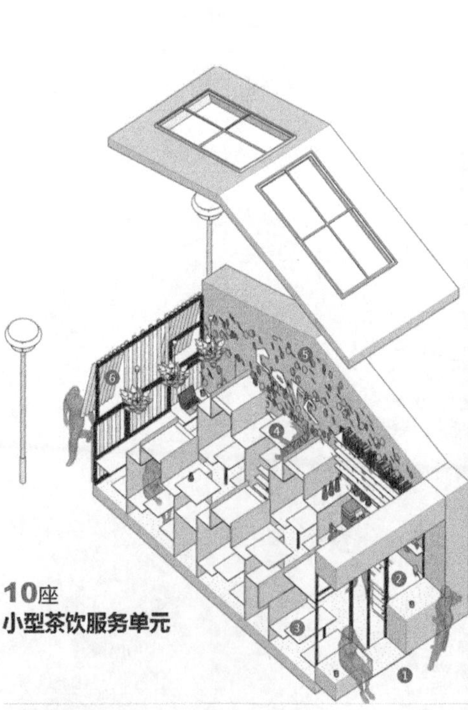

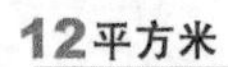

12平方米 → **10**座 **小型茶饮服务单元**

图16　多种人群背景的共享公寓与商业设施模型

（图片来源：自绘）

三个层级的共享,将历史建筑中零散而有限的空间组织起来:一方面将社群更新、文化展示与城市服务紧密结合;另一方面,将同时为原有居民部分改善居住条件、提升居住环境,并提供示范性自我改造或托管改造设计的可能性,达到原社区群体、新社区血液和旅游群体的利益与价值观共享(图 17)。

图 17 北京旧城现有空间模式与"共生北京"共享空间模式的对比

(图片来源:自绘)

结语:"共享生活"——一种可持续性、可替代性的生活方式

西欧各国的城市虽然有大量历史文化片区,但最终城市更新主体是落实在产权主手中,落实在个人手中,核心职能仍然是居住。

法源寺项目同样致力于合理利用现有条件,以个人为基本单元,以居住为核心,借助共享经济原则,提高整体社群的人居水平、服务水平与整体活力。

- 以现实的房屋单元为基础,借助设计与部分公属房屋与公共空间的便捷共享,提供基础的厨卫设施,是居民基础生活单元与人居尊严的前提。
- 以院落为单元,借助合理成套的居住单元,引入相对稳定的青年社群与家庭社群,形成与传统人居社群融汇共生的有机体。
- 融合新兴青年居民诉求的服务设施,整体提高社区的综合服务水平,并为文化遗产展示与活化、合理的文旅产业提供机遇。

要达到这一目标,有三个重要因素(图 18):

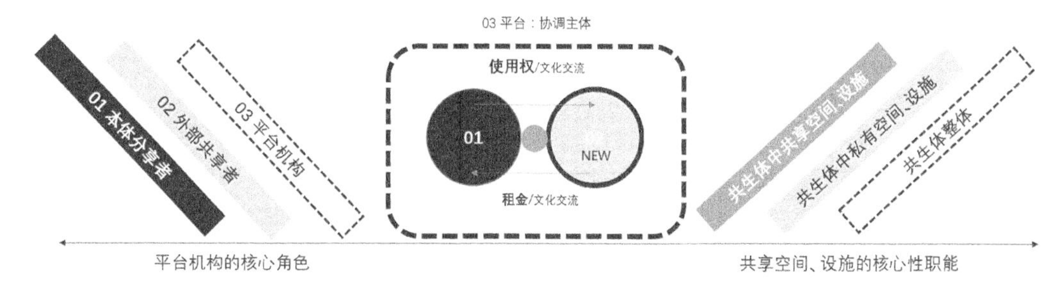

图 18 共享空间模式的要素

(图片来源:自绘)

• 原有居住者与新居住者的各自利益回馈与共同价值共识。

• 他们应分别拥有一个具有合理品质的私密人居环境，并借助一个优良的共享设施进一步提高其社会融合度与人居生活品质。

• 对此，一个具有包容性、开放性，并能进行精细化管理的整体平台，具有核心性作用。

在城市发展及更新过程中，合适的空间是“改变生活方式”“改变社会”的载体。

2015 年被称为共享经济的元年。“共享生活”的提出有其时代背景、环境压力和文化价值。经济放缓使得消费者接受获得产品和服务的使用权而不是所有权，“共享而不是占有”成为一种聪明的消费选择。在上述一系列以人居空间为核心共享载体的实验项目中，共享内容与频度的加强，与群体总量、多元性的增加，将使这一价值观真正成为城市人居文化领域的新型可能(图 19)。

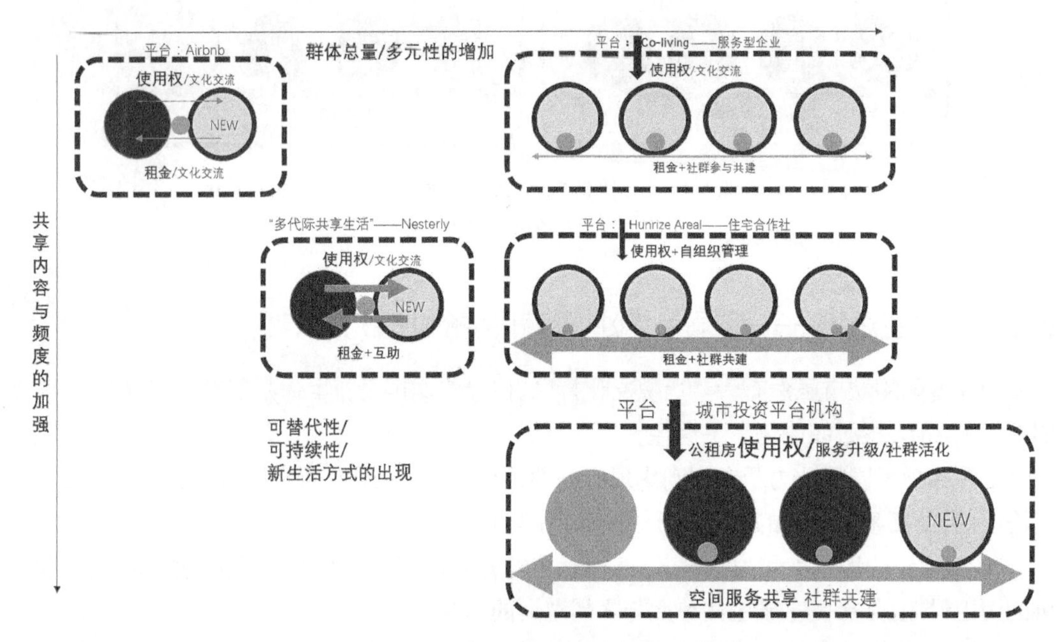

图 19　共享空间模式的模式对比

（图片来源：自绘）

整理公共空间、集约化组织私人空间的“共享生活”方式在国际大都市层面是一种新型的生活模式。在中国，它对于城市更新改造具有特殊意义。法源寺在旧城改造中关于共享理念的探索，提供了以集约化、功能化的空间改善老城生活最迫切的基础设施问题的可能，也为青年群体与传统社区之间提供了一种可持续、可替代的共生方式。

参考文献

[1] 徐勤政，何永，等.基于街区诊断的大栅栏地区城市更新路径研究[EB/OL].(2018-07-20). https://mp.weixin.qq.com/s? __biz=MzA3NTE1MjI5MA==&mid=2650768657&idx=1&sn=526c45a1c7ad78582a3561f3b19df726&chksm=877fcbdcb00842ca34d40dda37708bc12fd742ea51dce8c7486d88c2ba16c6d9235a124e8aaf&scene=27#wechat_redirect.

[2] 中国新闻周刊.北京租房报告：月薪一万，租不起房[EB/OL].(2018-08-24). http://finance.sina.

com.cn/china/dfjj/2018-08-24/doc-ihicsiaw3071789.shtml.

[3] 许欣欣,李培林,陈光金.社会蓝皮书:2015年中国社会形势分析与预测[M].北京:社会科学文献出版社,2014.

[4] Green G. The logistics of harmonious co-living: Exploring contemporary co-living through design interventions[D]. Kalmar: Linnaeus University, 2017.

[5] 蔡敏婕,姚瑶.美国爱彼迎计划在广州打造广府文化特色度假住宿区[EB/OL].(2019-04-03). https://www.sohu.com/a/305778893_100253795.

[6] Christensen, K, Levinson, D. Intentional communities and children[M]. Encyclopedia of community: From the village to the virtual world[M]. New York: SAGE Publications.

[7] 隰晓宇.基于共享生活模式的高层住宅设计研究[D].北京:北京交通大学,2017.

[8] Company, F. The Airbnb for affordable housing is here[EB/OL].(2017-11-21). https://www.fastcompany.com/90151804/the-airbnb-for-affordable-housing-is-here.

[9] 司马蕾.同一屋檐下的"共异体":老人与青年共享居住的可能性与实践[J].城市建筑,2016,22(4):28-31.

[10] 全球城市观察.上班族重返"集体宿舍",美国共享居住热潮[EB/OL].(2019-09-28). https://new.qq.com/omn/20190928/20190928A03DTV00.html

[11] 意厦规划设计(北京)有限公司.法源寺"共享街区"更新[R].北京:意厦规划设计(北京)有限公司,2019.

第四部分

作者简介

段　进　中国科学院院士、东南大学教授、全国工程勘察设计大师。

主要从事城市规划设计与理论的研究。创建了城市空间发展理论，提出“空间基因”并建构了解析与传承技术，较好地解决了当代城市建设中自然环境破坏和历史文化断裂的技术难题，并成功应用在雄安新区、长三角一体化示范区、苏州古城、南京2014青奥会等重大项目以及广泛的古城保护与新区建设中。研究成果被多部国家行业技术规定、指南、导则采用，曾获全国优秀规划设计一等奖5项、省部级科技进步奖一等奖2项、国际城市与区域规划师学会(ISOCARP)卓越设计奖、欧洲杰出建筑师论坛(LEAF)最佳城市设计奖等。

乔纳森·巴奈特(Jonathan Barnett)　宾夕法尼亚大学城市研究所研究员，宾夕法尼亚大学韦茨曼设计学院城市和区域规划系荣休实践教授。曾担任宾夕法尼亚大学城市设计项目主任。出版和发表了许多关于城市设计理论和实践的书籍和文章，包括最近的两部作品:《设计巨型区域》《重新发现开发法规》。

作为一名城市设计顾问和教育家，曾担任查尔斯顿(南卡罗来纳州)、克利夫兰、堪萨斯城、迈阿密、纳什维尔、纽约、诺福克、奥马哈和匹兹堡等美国城市，以及厦门和天津等中国城市的顾问。还曾任美国、韩国、巴西、澳大利亚多所大学的客座教授，并兼任东南大学客座教授。

乔纳森·巴奈特毕业于耶鲁大学建筑学院和剑桥大学。作为城市设计教育的先驱，他曾获得戴尔城市设计和区域规划卓越奖、新城市主义大会雅典娜奖章和威廉·H.怀特奖。他是美国建筑师协会和美国注册规划师协会的会员

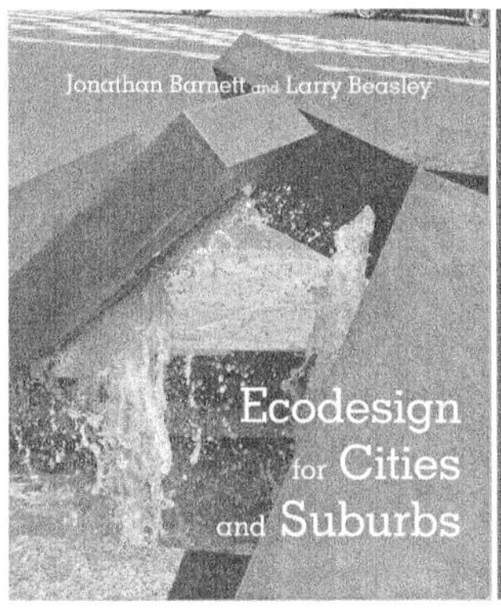

董亦楠　东南大学建筑学院至善博士后，意大利都灵理工大学访学博士，江苏省勘察设计行业协会城市建筑再生工作委员会秘书长。主要研究方向为城市建筑学、历史地段保护与再生，长期从事城市更新、历史地段保护再生的研究和设计工作。主持2项国家和省部级科研基金项目，曾获多项省部级各类设计奖。

郭　婧　北京市城市规划设计研究院城市设计所高级工程师、城市更新中心主任研究员，兼任雄安新区勘察设计协会城市设计分会委员。长期从事城市更新、城市设计与精细化治理等领域的规划工作。参与《北京城市总体规划(2016年—2035年)》《首都功能核心区控制性详细规划(街区层面)(2018—2035年)》等工作，作为项目负责人开展了《王府井商业区更新治理规划》《北京市城市设计导则》《北京国土空间规划城市设计编制技术要点》《北京城市公共环境艺术编制导则研究》《北京城市精细化管理机制研究》《北京街道更新治理城市设计导则》等课题，获全国优秀城乡规划设计成果一等奖2项、二等奖1项、三等奖1项。

韩冬青　东南大学建筑学院教授、博导，东南大学建筑设计研究院有限公司总建筑师、院长。主要从事城市建筑设计与理论的研究。倡导城市建筑系统观，强调建筑与场所环境及其历史文脉的密切关联，坚持专业品质与社会责任的兼顾，善于建筑创作与城市设计的互动融合。曾获全国优秀工程勘察设计行业奖一等奖2项、中国建筑学会建筑设计奖一等奖2项、教育部自然科学奖一等奖1项、国家级教学成果一等奖1项等。

安娜·尤莉安娜·海因里希（Anna Juliane Heinrich）博士，德国柏林工业大学城市与区域规划研究所“城市设计和城市发展”教研室研究员、讲师。研究重点是社会基础设施（尤其是教育基础设施在空间发展战略方法中的作用）、城市发展过程中的参与和共同创造，以及规划和设计中的研究方法论和技术。目前是柏林工业大学“1265‘空间再构想’”合作研究中心（Collaborative Research Centre 1265 “Re-Figuration of Spaces”）的成员，致力于重建儿童和青少年空间知识的变化。除了研究活动外，还在非营利组织“青年建筑城”（YOUTH ARCHITECTURE CITY）工作。

吕传廷　广州市城市规划编制研究中心原主任、中国城市规划学会理事、中国城市规划学会详细规划学术委员会主任委员、广州市人民政府第四届决策咨询专家。

主要从事城市规划管理工作，具备丰富的城乡规划编制、规划管理、公共政策研究经验以及国内一流的学术专业水平与政策研究能力。主持过多项国家、省、市级研究课题与重点城市规划项目，获得国际、国家、省、市级优秀规划设计奖几十项，著作及论文多次在核心期刊发表或国内外学术会议上宣讲。近年来主要负责广州市城市发展战略规划研究、“总控联动、多规合一”一张图，以及国际金融城、白云新城、黄埔临港经济区等多项重点工作，特别是在广州国际金融城开创了全面创新、岭南特色的城市管理机制。

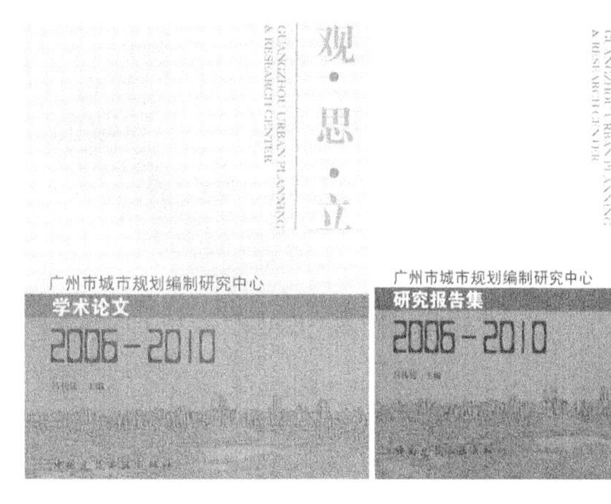

李　昊　西安建筑科技大学教授、陕西省教学名师、“城市设计”专业负责人。从事城市设计的教学与科研工作，专注于历史城市形态研究、地域性城市与建筑设计、城市更新与建筑改造研究。先后参加科技部国家重点研发计划重点项目“城市既有工业区更新与功能活化”等国家、省部级课题，主持完成“西安中轴线长安南路综合改造规划设计”等城市及建筑设计项目。获得国家教学成果奖3项。指导学生参加国际建协（UIA）大学生建筑设计等国内外重大设计竞赛，获奖50余项。出版“西安明城志：中国历史城市文化基因系列丛书”等著作8部，在国内外核心期刊上发表文章40余篇。

石晓冬　北京市规划和自然资源委员会党组成员，北京市城市规划设计研究院党委书记、院长，北京市规划展览馆馆长，教授级高级工程师；兼任首都区域空间规划研究北京市重点实验室副主任；新版北京城市总体规划及《首都功能核心区控制性详细规划》技术总负责人；获全国优秀科技工作者、国家百千万人才、国家有突出贡献中青年专家、国务院特殊津贴专家等称号；主持的重要规划项目获国家级优秀工程设计金奖1项、全国优秀规划设计一等奖8项、二等奖8项。

魏尔·阿尔陶克（Uwe Altrock）博士，城市规划师，德国卡塞尔大学城市更新与规划理论专业教授。他作为主编，出版了《德国城市更新年鉴》（系列）、《新欧盟成员国的空间规划和城市发展》（Ashgate 2006）、《成熟的特大城市：渐进式转型中的珠江三角洲》（Springer 2014）和《规划历史之窗》（Routledge 2018）等著作。他的兴趣和研究领域包括：城市治理、特大城市、城市更新与规划、规划理论、规划史。

杨震　重庆大学建筑城规学院教授、博士生导师，英国卡迪夫大学博士（城市设计方向），中国建筑学会城市设计分会理事，重庆市学术技术带头人后备人选（建筑学），重庆市首届“创新争先奖先进个人”称号获得者，重庆市首届“优秀青年建筑师”称号获得者。

主要研究领域为：山地城市设计、生态城市设计、城市更新、公共空间与社区营造等。近年来，在国内外核心期刊及国际会议上发表中、英文论文50余篇，出版英文专著2本（独著1、合著1）、中文专著3本（第一作者1、合著2），译著2本（合译），主持国家自然科学基金、“十三五”科技课题、省部级纵向科研5项，主持重大城市设计及城市更新类项目30余项，获得国家、省、部级以上奖励11项，包括全国优秀城乡规划设计二等奖1项、重庆市优秀城乡规划设计一等奖4项。

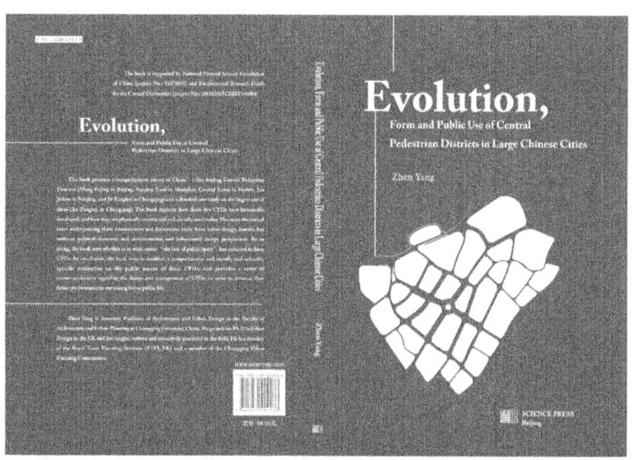

易　鑫　东南大学建筑学院副教授、硕导，东南大学中德城乡与建筑研究中心主任，欧洲规划院校联合会（AESOP）——欧洲和中国的城市转型学术委员会主任，中国城科会历史文化名城委员会数字名城学部秘书长。主要从事城市更新与历史保护、遗产数字化、城市设计相关领域的研究。主持自然科学基金2项，省部级课题多项，共出版著作8部（其中1部为德文独著），在国内外核心刊物上共发表期刊和会议论文40余篇，其中SSCI检索论文2篇。作为项目负责人，先后主持和参加了20余项工程项目设计，担任“城市愿景1910|2010：柏林·巴黎·伦敦·芝加哥·南京·北京·青岛·广州·上海”国际城市设计巡展创始策展人。

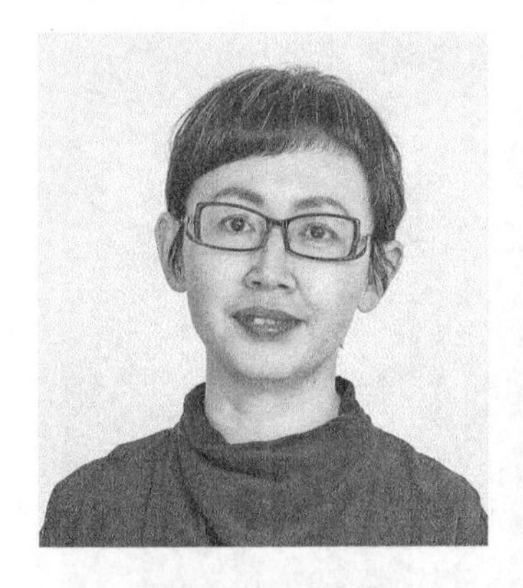

张亚津 德国斯图加特大学博士、北京交通大学兼职教授；德国注册规划师、建筑师；德国ISA意厦国际设计集团执行董事，北京、广州分公司总规划师。主要从事新城规划，以及空海港经济区规划、历史文化片区活化、科技创新等特定城市产业策划类相关研究。特别关注西欧城市发展与更新经验，在中国城市化进程中的对比性经验借鉴，并成功应用于北京副中心、广州空港经济区、广州临港商贸区、北京法源寺历史文化街区等项目中。曾获全国优秀规划设计二等奖2项、省级优秀规划设计奖项多项，德国ICON“标志”设计最高奖（Best of Best）、德国设计奖（German Design Award）银奖、美国“Green 100”等。